The Miracle of Touching

The Miracle of Touching

by
Dr. John R. Hornbrook

with
Dorothy Fanberg Bakker

HUNTINGTON HOUSE INC.

Shreveport ● Lafayette
Louisiana

Copyright © 1985 by John R. Hornbrook
ISBN Number 0-910311-28-5
Library of Congress Card Number 85-80004
Printed in the United States of America

DEDICATION

This book is dedicated to my lovely wife, Colene, who touched my life and turned it into something beautiful. Until she touched me, I never understood the meaning of love. It is further dedicated to Robbie and Shellee who taught me the meaning of the unsurpassed love that only children dare to demonstrate. They taught me how to touch others through their need to touch and be touched in return.

ACKNOWLEDGEMENTS

I would personally like to thank my good friends Norman and Dorothy Bakker who have been sources of encouragement for nearly five years. I am so thankful to have friends like these who not only have touched my life, but the lives of so many others with whom they have come into contact.

It would be impossible to personally thank all the people that have left a lasting impression on my life. However, I would like to thank a few friends who have touched me through their contacts. Special thanks are extended to Pat Boone, Efrem Zimbalist, Jr., Connie Smith, and all the staff at the PTL Club who treated me with kindness and encouragement. Also, I will be ever grateful to my dear friend and pastor, Steve Wyatt, along with all the folks of the "Cullen Avenue Church Family" for their love and support.

Lastly, deep appreciation is extended to my many special friends at Nova University who influenced my decision to further my training in the area of Early Childhood. Special thanks are extended to Dr. Diane Marcus who encouraged me to keep writing to help others and to my compassionate professors, Dr. Mickey Segal and Dr. Jeri Sorosky, who taught me how to influentially touch the lives of young children.

INTRODUCTION

When I was small, my mother was afflicted with a chronic long-term illness, incomprehensible to a little boy. She suffered from severe joint destructive rheumatoid arthritis. As a consequence of this illness, I was deprived of the frequent tender touches of a loving mother that an insecure lad requires for his growth and development. Although she breast-fed each of her children for a perfect start in this life, her illness took her away from me much too early.

Whether by plan, or accident, or by divine intervention, I was shipped out to my beloved maternal grandmother for her care and guidance. I owe her very much for my security and development of the social graces that are so vital in interpersonal relationships today.

Everything about me was, to a degree, molded by her. My early schooling was at her knee. Her guidance of me was mostly by her tender touch.

She was a school teacher and her hand was a correcting instrument when she was displeased. She wore a sewing thimble on the middle finger of her left hand and, with that small instrument she could effect immediate corrective behavior with a small, and inconspicuous (to others), thump behind one of my ears.

When Dorothy Bakker asked me to express my opinion of Dr. Hornbrook's new book, *The Miracle of Touching,* I was delighted.

The manuscript was so absorbing that I couldn't put it down but read it through in one sitting. The thoughts expressed here are the written embodiment of my early childhood as well as a very basic tenet of my medical practice these past 40 years.

During the clinical years of my medical education, I noticed that the professors with the highest recovery rate and shortest hospital stay were those individuals who had the best rapport, and the warmest relationship with their (our) patients. Those doctors seemed always to reach out and touch their patients. These patients invariably brightened when the professor came into the room at the prospect of his unique "bedside manner."

The same professors were brutal in their discipline of us in our bumbling, stumbling entrance in the world of service to the ill and injured. The inept were just not tolerated. For this I am eternally grateful, as it put me "on point" as it were against the inroads of Satan as he fastens himself onto the sick, injured and dying. A physician with a "physician's heart" is surely and definitely at war against Satan all of his life.

While studying in Vienna, Austria, I saw recently-operated patients, severely injured patients and even leg amputees attempt to stand up out of deep respect and in anticipation of the professor's touch.

Throughout my medical career I have utilized the tender touch to draw closer to my patient, holding his hand, or his foot, touching him on the back, or cheek, wherever was least painful to him. This technique has always been extremely useful if the situation required me to deliver bad or devastating news. I always touched a family member whenever required to impart any information about a loved one. It was never resented or withdrawn from but was always eagerly accepted.

At one time I worked closely with an extremely brusk, remarkably skilled and capable surgeon. His bedside manner was atrocious and curt. It came to the point that he asked me to do all of the pre- and post-operative discussions with his patients as he just did not seem to be able to adequately communicate with them. His problem was simple. He was aloof and cold and never touched his patient, except to examine him, or with a scalpel on the operating table.

To quote from Dr. Hornbrook in his discussion on the effects of touching: "To the lonely, unloved or sickly — even the comatose or dying — the touch can be the essential ingredient needed to accomplish a miracle. The human touch can revive a will to live which has wilted like the petals of a fallen rose. In some cases, almost magically, the deep subconscious slumber of a coma has been penetrated by the touch of one who cared. Thus the touch has become the 'miracle cure' for an ailing body, will and soul.

"Jesus, our blessed example, was a toucher. He touched the young and the old, the blind and the sighted, the rich and the poor, the clean and the dirty, the respected and the despised, the saint and the sinner and literally everyone He ever met."

This small monograph should be required reading for every pastor, elder, deacon, physician, intern, counselor, and social worker, as well as every parent. In fact, it should be read by everyone confronted by interpersonal relationships.

This flowing, lucid interpretation of one of the basic principles of the Christian walk, "The laying on of hands," has allowed me once again to feel the tender touch of my grandmother (also her thimble). It allowed me to relive the joy of participation in God's healing of the sick and dying, as well as the exhilaration of being a part of the birthing miracle of new life into this world.

Lester E. Nichols, M.D.
Palmdale, California

TABLE OF CONTENTS

FOREWORD

Have you ever wondered, "How can I be of great value in our turbulent world?" Perhaps you have asked yourself, "What can I contribute to bring eternal satisfaction to the inner void within my being?" Many have sought to leave their mark upon the world, but few have succeeded.

Throughout this book, we will explore avenues by which we can all utilize our unique abilities to leave lasting imprints upon the sands of time. Each person is different but of vital importance to others as individuals and to the world as a whole. You and I do make a difference! As we journey together, we will see just how significant our lives really are. We will also explore how our individual variances can be used to bring happiness to both ourselves and those around us.

Life is a gift to be enjoyed — not merely an existence to be endured or tolerated. It is possible to experience happiness, security, success, health and wealth — but only if we will dare come out of our shells and into life's arena. We are actors on life's vast stage and the success of the production depends on our willingness to put our whole heart into the play before the final curtain call. Then we can take our second bows in eternity before the celestial hosts and redeemed multitudes!

Happiness depends upon loving and being loved and this book was written to help you, and others like you, discover how to express feelings of affection. Not only is the expression of feelings explored, but also the development of positive self-images.

Learning to like people is the key to the door of ultimate success. Together we will learn to practice the art of winning friends and liking them, as well as learning to love ourselves. Loving one's self is not only natural, but psychologically healthy unless carried to extremes. Stable people can care for others because they like themselves and are willing to correct personal character flaws.

Throughout our journey together, we will view touching as synonomous with reaching the inner soul or heart of others. Both physical and emotional expressions of emotions are vehicles by which this may be accomplished. True touching is that which channels all the way to the depths of man's mind, body and soul. The ultimate touch permeates mankind's entirety, touching the physical, emotional and spiritual.

We have no greater gift to share than our touch. It is the ultimate form of expression by which communication is accomplished. This priceless commodity could be described as an art form to be presented by you and me as life's artists. Won't you take the time to develop into one of life's immortal touchables? Come travel with us and we will learn how to become among the most loved people in our communities, while bringing joy to others around the universe.

John R. Hornbrook

1

EFFECTS OF TOUCHING

To the lonely, unloved or sickly — even the comatose or dying — the touch can be the essential ingredient needed to accomplish a miracle. The human touch can revive a will to live which has wilted like the petals of a fallen rose. In some cases, almost magically, the deep subconscious slumber of a coma has been penetrated by the touch of one who cared. Thus, the touch has become the "miracle cure" for an ailing body, will and soul.

A caring touch can turn lives around which have strayed from acceptable conduct. Researchers in the fields related to social behavior have discovered that when adolescents are touched by warm, smiling, friendly teachers, academic performance levels increase and school dropout rates decrease.

Pro-social behavior is directly influenced by the manifestation of love and warmth from a model who is willing to be affectionate and show inner feelings. The touch is a positive reinforcing agent which can stimulate desired emotional, physical and spiritual levels of attainment.

Research has proven that children who are touched physically and emotionally by an adult authority figure are more considerate, altruistic, loving and sharing. Granted, this is a bold statement — but it is scientifically based upon the results of valid laboratory tests by social psychologists.

In one study, children were given the choice of receiving candy or the affectionate response of the experimenter. This was done in the following manner: The child sat close to the experimenter who demonstrated the operation of a "choice box" which had two levers. If the experimenter pressed one lever, he received candy. If the other lever was pressed, a flashing red light was activated. The experimenter demonstrated no emotional response when candy was received from the box, but each time the flashing red light was activated, she responded in a kind tone, smiled and hugged the child. Repeating this process, the child's emotional reinforcement (triggered by the hugs, and kind responses) became associated with the experimenter's response in choosing the flashing red light.

Later, each child had an opportunity to operate the box while the experimenter sat across the table from the child. One choice would render candy, while the other would bring affectionate approval from the warm, touching, caring experimenter. If the child chose one lever, the physical reward of candy would be received; however, the other lever would bring the delight of the experimenter's reward of emotional gratification.

The majority of the children chose the flashing red light, which brought the smiles, hugs and affection of the experimenter. When children will choose the touch over candy, the value of touching is beyond dispute.

Psychoanalytic theorists believe that empathy for others begins developing during the earliest months of infancy — caretaker interactions between infants and their parents or whoever cares for their needs during infancy. The belief is maintained that moods are communicated to infants through touch, tone of voice and facial expressions.

If an infant is caressed, spoken gently to and smiled at frequently, the infant becomes more loving as he matures and possesses a quality of concern for the welfare and feeling of others.

Thus, again, touching teaches touching from the onset of life.

A lifestyle is modeled that is incorporated for the betterment of humanity, as well as the person learning to empathize. Parents, grandparents, babysitters and others who are a part of an infant's life, should be careful that the need to be touched, smiled at and softly spoken to is not neglected. This is not a suggestion, but a mandate for the emotional development of the young.

In most societies, cultures are comprised of touchers and non-touchers alike because of personality differences and the varied experiences of each individual's life. Simply stated, *most* cultures are neither characterized as being made up of people who touch — or of people who refuse to touch and be touched.

Relevant to proving the importance of the touch, I would like to submit two exceptions to the rule (that most societies are a mixture of both touchers and non-touchers). These two examples demonstrate how the extended touch (or absence of it) can influence an entire culture.

First, let's take a look at the Ik, the mountain people of Uganda, as described by anthropologist Colin Turnbull in *The Mountain People*. This small tribe of hunters once had a social and moral culture with mores, laws and civilized customs — until they were deprived of their hunting grounds. They then became divided into small groups of renegades, whose only concern was survival.

The Ik became untouchable and uncivilized, characterized by treachery, lying, stealing, deceit, killing and savage behavior. No longer did the people show love and compassion — not even to closest family members. Feelings of caring, mercy and empathy no longer existed. All personal human contact was hostile — except for mating, which was an unemotional act solely for the purpose of procreation of the tribe. Civilized behavior had totally vanished with the abolishment of the humanitarian touch. Isn't there a lesson to be learned from the Ik?

Secondly, let's look at a culture which was the complete opposite of the Ik. It is the Hopi Indian tribe of Arizona, which became known for its emphasis on developing a "Hopi good heart." A "Hopi good heart" meant trusting others, respecting all people, being concerned for "everyone's" rights, having an inner peace, and avoiding conflicts with others. Hopi mothers cuddled their babies in a loving manner.

Mutual respect and cooperation was a way of life in the Hopi family as well as in the tribe as a whole. The society could be characterized as a loving, helping, sharing and "touching" society. This small society manifested the touch being extended "to everyone by everyone" through love and "respect for all by all."

What if modern societies would pattern themselves after the Hopi? More specifically, what if all Christians were to apply the teachings of Jesus, which are synonomous to those of the Hopi, regarding the treatment of one's fellow man? Oh, how the world could be changed by people who sought a "Christian good heart." People who would reach out to the perishing with open arms of acceptance and love's tender transforming touch.

Almost identical were the findings of Margaret Mead, as related in *Sex and Temperament in Three Primitive Societies.* She found a tribe of touchers and a tribe of non-tactile natives living on the same island in New Guinea. Members of one tribe, the Arapesh, were gentle, loving, tactile and unaggressive and were concerned about the needs of others. Feelings of others were highly regarded and efforts to display warmth were exerted. Children were respected and cared for lovingly with much physical embracing.

In contrast to the Arapesh were the Mundugamor men and women. They were ruthless, quarrelsome, uncooperative, non-tactile and held little regard for the lives and property of others. The Mundugamor children were abused physically, and embracing was viewed as a sign of weakness in young and old alike. A society void of human physical contact had developed an angry, abusive clan.

I interpret these studies to prove without reservation that the kind touch produces fruits of warmth, empathy, generosity, concern for others and a positive self-image. Likewise, the rejection of any type of tenderness produces hostilities, isolation and anti-social behavior. "Hands on" people are touched in return, while the "hands off" crowd goes virtually untouched by others. What one hands out is what one will receive in return.

In another chapter, *"A Point Of Contact,"* the application of a *faith-released, spiritual* touch will be discussed. This touch is similar, yet different from the mere physical touch now being practiced by medical science. Both techniques have recorded striking results from the "laying on of hands."

In a scientific approach to the phenomenon of "laying on of hands," nurses are being taught the practice of touching patients to alleviate anxieties, reduce pain and stimulate healing.

By touching patients lightly, this "therapeutic touch" has enabled nurses to ease the agony of backaches, reduce the pain of women in labor, and speed up the healing process. One leading physician related that he had no doubts that the therapeutic touch facilitates healing. Medical science cannot explain how it works, but in his judgment, it does have validity.

According to a leader in the field of nursing, this practice is now being taught at more than fifty universities and hospitals. Thousands of nurses have been instructed in this technique which is now being practiced nationwide in clinics, hospitals and nursing homes.

This technique is quite simple, but very productive. A nurse practicing the therapeutic touch places her hands close to the patient's body, then lightly touches the patient's body from head to toe, while reassuring the patient's confidence.

This treatment has been very successful in treating heart patients by decreasing anxieties, and has helped to decrease the nausea and vomiting side effects associated with cancer therapy. A growing number in the medical profession view the therapeutic touch as a valid scientific tool, worthy of implementation.

As reiterated throughout this book, the human touch possesses wondrous results, whether in social, medical or spiritual realms.

Perhaps the greatest single quality transmitted by the touch is the transference of love from one person to another. Love is the most powerful force in the world and is capable of healing broken minds, bodies and spirits. The touch is the vehicle by which love can be conveyed.

An entire book could be written demonstrating the positive influences and effects produced from research focusing upon interpersonal tactility. However, that is not the focus of this book. The intention of this chapter is to present a scientific foundation for establishing credence in regards to the effect of the touch.

The various informational blocks relating to touching will be built upon this foundation as proven through valid research, along with other reliable sources and experts worthy of acceptance.

2

REACH OUT
AND TOUCH SOMEONE

In 1980, the Bell Telephone Company first introduced the popular ad which touched the heartstrings with the slogan, "Reach out ... reach out and touch someone!" Today there are millions of desperate, lonely people longing to be touched by someone who cares. Many need so desperately to be emotionally embraced that they turn to substitutes in the form of drugs, alcohol, excessive work and sex. The need for living is often lessened by the trampling feet of loneliness, which accounts for the rapidly increasing suicide rate.

The late Hank Williams wrote lyrics describing the legacy of loneliness. "Hear that midnight train winding low, I'm so lonesome I could cry."

At the age of twenty-nine, this musical genius died all alone in the back seat of a car, half way to his next show.

The deadly fangs of loneliness are no respecter of persons. They rip the rich and poor alike. Position, background, social status, age, race or religion make no difference; everyone needs to be loved. EVERYONE needs to be TOUCHED ... to know there is at least one person who cares about them and needs them.

The eccentric tycoon may publically boast that he's a self-made man. "I made it to the top all by myself," he boasts. "I don't need *anyone* or *anything!*"

He would never admit his talk was just a cover-up to conceal the emptiness of his soul. No mention is made of the countless nights when the silence is broken by the pathetic sobs of the lonely tycoon crying himself to sleep.

The glamorous life portrayed by Hollywood is a misconception causing a plastic-world effect which deceives man. Beauty, fame and fortune are depicted as ultimate forms of happiness, often elevating their possessors to god-like positions. If things could bring true happiness, why did sex-symbol Marilyn Monroe die all alone of a sleeping pill overdose? Why did the promising young actor Freddie Prinze shoot himself at the height of his career? Why did Howard Hughes die a reclusive, self-imprisoned, needle-riddled, lonely shell in spite of all his billions?

Beauty, fame and fortune are not capable of filling the void in the human heart which demands the loving, caring touch of another person. The gratification of success cannot satisfy the innate nature designed to flourish in love.

Our civilization has made great scientific advances in the field of prosthetic appliances. Man has been able to produce artificial limbs, bone replacements, sight and sound devices and heart bypass valves. He has also made great strides toward perfecting a functioning artificial heart. But one thing he will never be able to duplicate is the inner emotions of the heart. Science has developed medical cures for thousands of ailments but has never been able to produce a "love pill" to alleviate the agonizing pain of loneliness.

Man has certainly tried! He has developed the sleeping pill to help people get through the long night hours, only to face their loneliness anew the next day. He has developed tranquilizers to help mask the frustration of unfulfilled needs, but it never provides a cure.

Remember, you can put saccharin in your tea and milk-mate in your coffee — but there will never be an effective substitute for love in your life!

The need for touching was so aptly demonstrated by the Coca-Cola Company's award-winning commercial of 1980. The episode showed a little boy timidly handing "Mean Joe

Green" a bottle of Coke after a grueling football game. "Mean Joe's" countenance turned from one of stress to a warm smile as he tossed his jersey to the boy who had given him the Coke. The commercial was rated number one because Joe showed compassion by reaching out to his young fan. Viewers related to the incident because they too have a deep-seated longing to be emotionally embraced by others.

Celebrities are often mobbed by affectionate fans clamoring for autographs, mementoes or a physical touch. Great satisfaction and fulfill-ment of the need for touching is derived from being close enough to touch the celebrity in a person. An autograph is a trophy proving the adventurous encounter.

One day after I had been a guest on the PTL Club, several of us, including my friend, Pat Boone, were eating lunch at the ministry's restaurant. I left to go to the restroom when I was cornered by two elderly ladies asking for my autograph and wanting to know how they could also get Pat's. Seeing the longing for an en-counter which would bring ultimate joy to two little elderly ladies, I said, "Hold on just a moment and I'll get Pat's autograph for you." I returned moments later with a personal note from Pat to each of the ladies and saw the elation on their faces as they each received a treasured possession; a touch from a celebrity who cared enough to reach out to others.

I have never been an autograph seeker and was surprised when people started not only asking but demanding autographed copies of my first book when it was released. Then the Lord showed me there was a ministry of love in reaching out to touch others through a personal notation in a book. I had promised God long before He opened the doors to share through books and television that regardless of how far He took my career, I would never forget to be "close" to people.

My greatest joy has been derived from mingling with people... touching them. After doing television shows with live audiences, such as PTL or 100 Huntley Street, the thing I look forward to most is meeting the people who have come to the program and praying for their personal needs.

My friend "Uncle Henry" Harrison who is seen regularly on PTL, spends much of his time loving the people who come to be a part of the ministry. He is an ambassador of mercy with God's highest proclamation, reaching out and touching people with love. It is God's love which flows through Henry to touch others.

A lovable, cuddly, godly "teddy bear" is what my wife calls Henry. He is not ashamed to show emotions, nor is he too big to be humble. The personal touch of this spiritual giant has directed countless individuals to the loving arms of Jesus through supernatural love.

The universal need of humanity is so aptly stated in the lyrics of a song from the '60s; "What the world needs now is love, sweet love." God needs more people who will love the unlovable and touch the untouchable with the love of Jesus. *You* could be the one!

Beloved, won't you try this as well? Granted, it may seem awkward at first, but what a blessing you will be as God pours out His blessings on you and others because of your obedience! Don't be afraid to touch others. The touch is man's ultimate expression of personal concern and the vehicle by which he relates his feelings to those with whom he comes in contact.

Jesus, our blessed example, was a *toucher.* He touched the young and old, the blind and the sighted, the rich and the poor, the clean and the dirty, the respected and the despised, the saint and the sinner and literally everyone He ever met.

His compassionate touch knew no boundaries and ours shouldn't either. If God Himself is no respecter of persons (Acts 10:34), who is mortal man to exalt himself to the office of judgeship? That office is reserved for God alone!

3

AMONG THE
HIGHEST CALLINGS

Each person has a unique calling upon his or her life enabling fruitfulness in their special field of service. One may be called to reach the down and outs, while another is to reach the up and outs. "High Adventure" host, George Otis, former executive with the Lear Jet Corporation, was called to witness to the jet set of Hollywood. Obedience to his specific call enabled him to lead Pat Boone into the fullness of the Spirit. Indirectly, George has touched millions through Pat's ministry which continues to expand world-wide. One never knows the impact a touch will have when reaching out to another. Every born again Christian is indebted to someone who cared enough to tell him about Jesus. This favor can only be returned by passing the "good news" along to our unsaved brothers and sisters.

I believe God has escrow-reward accounts waiting in heaven for what we do on earth. Can you imagine how much interest is accumulated for the touchers who win great soul winners to Christ? One prime example would be Jim Bakker's grandmother who touched him by her love for Jesus. In return, her account has been credited for the millions touched by Jim Bakker through PTL. Each great ministry has resulted from its founder being touched along life's pathway by someone who pointed them to Jesus.

I am confident that when Pat Robertson met with a whiskey-drinking disc jockey several years ago he did not realize the importance of his touch. Pat reached out to a desperate man, leading him to Jesus. This man would later become a co-host to both Pat and Jim Bakker. Yes, I'm talking about the beloved "second fiddler," Uncle Henry Harrison. Henry was lonely, frustrated and searching for meaning to life which he found through the touch of a minister he didn't even know. Now both men are known 'round the world through television's outreach as they touch the lives of millions for Jesus. Now do you see how important a touch can be?

I Corinthians 13 lists the various gifts entrusted to believers within the framework of fellowship. Included in the nine spirituals gifts is the gift of "helps." I maintain that touching is a gift of helps. This gift is manifested through a genuine concern for others. When we touch, we become an extension of the Spirit who is love.

The Hallmark Company has captured the essence of touching by recommending their greeting cards to those who "care enough to send the very best." It seems, however, that caring in the sense of personal involvement has all but faded into obsolescence. In the face of today's crime, violence and moral decay, even Christians have joined the ranks of the majority by refusing to become "involved." With so many hurting and so few caring, it places enormous responsibility on those who have this highest of callings. It's up to you and me to change this shameful situation by openly reaching, touching and loving. If you were in need, wouldn't you welcome a friendly encouraging touch?

For nearly two decades, I have worked with people who are hurting. From skid row to the mansions of the elite on Snob Hill, all people have two things in common: the need to be loved and the desire to be touched. Even the hardened murderer on death row has a tender spot in his heart crying out for love. The hooker walking the streets, selling her body to the highest bidder, secretly longs for acceptance even though her activities bring her social rejection. An addict searching for the next "fix" needs a substitute in the form of a needle or pills to momentarily cause the aching loneliness to go away. The lonely wife of an executive sits forlornly amidst the plush luxury of her mansion, sipping vodka until the need to be touched is forgotten by the alcohol's numbing effect. Again I proclaim, everyone everywhere needs to be touched.

In 1979 we were driving to Charlotte to be on the PTL Club. Since Colene's relatives live in Morristown, Tennessee, we stopped there to visit for a few days. One day while Colene went to visit her uncle, my son Robbie and I were enjoying a swim at the Holiday Inn where we were staying when I caught a glimpse of a large bus in the hotel's circular drive. My heart fairly leaped as I realized the bus belonged to a very famous singer who had reached the number one spot on the charts more than once during his career. I felt an aura of excitement as thoughts of witnessing to this celebrity raced through my mind. Never before had an opportunity availed for me to witness to anyone in show business.

Still thinking how wonderful it would be to tell this person about Jesus, Robbie and I made our way into the motel restaurant to get something to drink. There, waiting to be seated was the entertainer along with members of his band. Timidity vanished, replaced by a strong compulsion to reach out and touch this man's life with Jesus' love. Since he was standing directly in front of me, I jokingly asked him "Do you know so and so?" mentioning his name. This made an impression with him, since he was dressed somewhat incognito. He shook my hand and we got acquainted, chatting mostly small talk.

I marveled at the warmth and friendliness of this man who had achieved such acclaim. Yet, a cloud of loneliness seemed to overshadow his smile. His eyes reflected a need to be loved; not for his fame and fortune, but just for being himself. This void became more evident as he talked about how he enjoyed charity work because it gave him opportunity to help others. Noticing the lavish array of exotic jewelry he was wearing, I thought, "This poor, poor rich man!"

Before getting a chance to tell him about Jesus, he was called to a reserved, private-seating area away from the regular customers. Momentary disappointment was dispelled by the gentle voice of the Spirit saying, "Go back to your room and inscribe a copy of your book with the inscription I shall tell you, and take the book to the celebrity." We rushed back to the room and jotted down a beautiful inscription, which I will not share in order to protect the identity of the individual. Then returning to the restaurant, Robbie and I walked boldly to where the private entourage was seated. "I hate to bother you," I explained, "but Jesus told me to give you this book." Much to my surprise, the man opened the cover and read the inscription. With tear-dampened eyes, he rose to his feet, put his arm around my side and gave me a gentle hug. Then with a quivering voice, he softly replied, "Man, that's where it's all at."

The mission having been completed, I wished him well while he stood with eyes fixed on the picture of Jesus on the book's cover.

The expression on his face when he had said, "Man, that's where it's all at," had broken my heart. He had looked like a prisoner gazing wistfully out the window to freedom while being locked up inside. Within, there was a void crying out for Jesus' touch while on the outside was self, clinging to the world's ways.

An inner war had been apparent, but perhaps one touch from a stranger helped him find the victory and peace which had so long eluded him. I drew comfort in knowing that one soul who had greatly needed to feel someone's loving touch, had been reached. Nothing leaves greater impact than the touch which is ordained by God.

A popular chorus voices the words, "O, to be His hand extended, reaching out to the oppressed. Let me touch Him, let me touch Jesus, so that others may know and be blessed." You and I are the only hands God has through which He can reach people. What happens if we fail to yield them to His use? Only eternity will tell!

4

A UNIVERSAL NEED

One year I served as a counselor assigned to handle discipline in an inner city school. On the day prior to a four-day break, I decided to begin growing a moustache.

Upon returning to cafeteria duty, I found several first- and second-grade students who could not believe that anyone could grow a moustache in that short a time. On that first day back at school, at least a dozen brave little troopers asked to feel my moustache, to make sure it was real. As I gladly accommodated them, I became aware of a new closeness with the students, brought about by the situation. They felt closer to me because the school disciplinarian was not "untouchable," but allowed himself to be human and within reach. A friend would permit touching, while a "cold" authoritative figure would keep a distance between himself and the ones entrusted to his jurisdiction and control. Beloved, no words can adequately describe the rewards of touching.

In order for me to more fully comprehend the essence of the human touch, the good Lord had to paint pictures for me from life's vast canvas. I think of the memory as God's divine "replay system" by which the past may be reviewed. Isn't God smart? He has placed within every human brain a system by which recall can be utilized. When we use the brain to either catalog or recall information, we are employing the greatest, most complex computer ever invented!

As the Spirit moved upon me to write, he reminded me of multiple episodes from the past.

In one picture, a television show with Dr. Lester Sumrall was replayed in my mind. Dr. Sumrall and I were discussing my book, *You Are Somebody Special,* and I shared how uniquely special we are as individuals. While reliving this exciting one-hour interview, God said, "Now notice your hands." Unrealized before was the touch being personified before my eyes like swaying trees upon a hillside on a placid summer morning. The Spirit bade "Watch! Watch! See how many times you touch brother

Sumrall when answering his questions." Sure enough, there it was; the communicating touch. Without realizing it, I reached out and patted the arm of Dr. Sumrall numerous times while making emphatic points during the interview.

Then my mind began to span the rivers of time, re-exploring meadows of the past. Crystal clear was the emphatic relevance of the human touch. Monuments of expression marched forward as though on review for spectatorship. The monuments were hugs, kisses, handshakes, winks, smiles and gentle expressions of tender loving care. Each gesture was symbolic of a bridge whereby one individual could cross to contact others, thus touching their lives and establishing bonds of love.

Gestures of friendship differ from country to country and even region to region. However, the uplifted or extended hand has become the universal symbol of peace and friendship. This symbol signifies the willingness of one party to touch the other party with tranquility rather than hostility or aggression. Statues of interclasped hands are found in numerous civilizations representing the brotherhood of individuals, clans, or mini-societies. Pictures depicting the significance of the human touch have been recorded by ancient man on the walls of caves spanning eons of time. The touch is truly universal, though varied in interpretation and expression.

Almost anyone who has watched any western

movies or television shows can relate to the sign of the uplifted hand of the Indian chief who inevitably says, "How." Yet, if the Indian considers the paleface worthy of paramount friendship, the union of blood brotherhood can be established. This is initiated by an incision being made on the hand of each party whereby the two can touch, allowing their blood to mingle. Not only has the physical touch related friendship, but also the life-sustaining blood. An eternal pledge of kinship has been ceremonially displayed for the world to see.

Jesus does the same for us when we become His "blood brothers." By accepting Him as Savior, His shed blood makes us His brothers. At this point, we become joint-heirs with Him. By reaching out to Jesus, we are granted the same privileges, power and honor as Jesus because we are legally part of God's family through our brother, Jesus. This adoption process is described in the eighth chapter of Romans and should be reviewed until it is fully understood. Christians must realize who they are if they are to live the joyous, victorious, mountain-conquering life which God intended. By realizing how important you are through Christ, you can claim your rightful position as a leader in God's brigade which is to reach the universal ache with a healing salve. This ointment for aching hearts is love's tender touch.

Everyone needs to be needed, and at least

subconsciously, desires to be loved. I have heard
hardened criminals boast, "I don't need any-
body or anything. Baby, I am my own main man."
This is merely a cover-up for the loneliness inside. Man was created with an inner void which
can only be filled with the love of outside forces.
This void has two empty chambers waiting to be
filled. One chamber requires natural love while
the other demands supernatural love. One was
designed to receive the love of humans while
the other seeks God's perfect love.

5

LAWS OF TOUCHING

Just as God's law of giving applies to our finances, it invariably holds true in the giving of one's self, or "touching." This principle, simply stated is: "He who touches shall be touched more than he is touched." The more one is willing to share, the greater the yield of multiplied blessings. It is impossible to outgive the Heavenly Father, outlove Jesus, or outtouch the Holy Spirit.

Each time a believer responds to the perfect will of the Master by loving others, outpourings of love are received through the personage of the Comforter, the Holy Spirit. The result is an unbroken cycle of love from the Holy Spirit, to us, to others.

In the words of a simple mountain man who had gained supernatural wisdom, comes a beautiful description of giving and receiving. "When you give someone a dipper of water in Jesus' name," he explains, "God refills your pitcher with a bigger dipper, causing your pitcher to overflow." This says it all, and I would not attempt to improve upon this gem of truth.

The truly happy individual is one who enjoys people, loves God and appreciates life. The Living Bible clearly defines the three steps one must take to attain this. It declares, "... you must work hard to be good, and even that is not enough. For then you must learn to know God better and discover what He wants you to do. Next, learn to put aside your own desires so that you will become patient and godly, gladly letting God have His own way with you. This will make possible the next step, which for you is to enjoy other people and to like them, and finally you will grow to love them deeply" (2 Peter 1:5-7). In order to find favor with God, we must be willing to share our very selves with others which is the ultimate of servility.

Before we can receive divine exultation, we must clothe ourselves with the humility of spirit that enables us to touch others. Through haughtiness, arrogance, pride and self-righteousness, many believers are self-restricted from touching and reaching out to those who need encouragement. Pride is a barricade which restricts intimate contact. It serves as a heat barrier, keeping the warmth of the loving, caring touch shielded. Pride carries as its counterpart a chilling atmosphere as cold as ice. This is why a humble, caring person is often referred to as tender, while the haughty individual is called calloused or snobbish.

The point I am trying to target is that Christians cannot look down their noses at those considered less fortunate and expect to effectually reach them. Nothing turns an unbeliever off more quickly than a "holier than thou" attitude from someone claiming to be a saint of God. While such a condescending attitude may be due to ignorance, the results can be devastating. Many such individuals never get a chance to make amends for their miserable first impressions because the one they offended has lost all confidence in them and wouldn't even give them the time of day. Many opportunities are forever forfeited because of a bad first impression. Trying to touch a person who has been offended in this way is like trying to pet a would-be vicious dog who has just been kicked in the face. Take my word for it and save your skin, it just won't work!

The reason many born again ex-convicts have successful prison ministries is because they are not afraid to touch those rejected by society. The inmates can both indentify with and respect someone who has been in their shoes. And understandably too, there are many laymen and clergymen who are totally ineffective in prison ministry because of their lack of sensitivity and humility. Prison life tends to make one a good judge of character. I'm told that the inmates generally size a person up and conclude whether he genuinely cares, or if he's merely doing his "good deed" for the week. If they feel it's the latter, chances are he'll never win any one of them to the Lord.

Let us continue on the more pleasant and positive aspects of the laws of touching. I feel, however, that this warning had to be issued, since touching others is of supreme importance to God, others and to ourselves as individual ambassadors representing heaven's embassy. Please permit me to paraphrase Jesus' teaching by the following declaration. "It is more blessed to touch than to be touched." Touching, like giving, brings more satisfaction and personal gratification than does receiving. The ultimate joy is derived from reaching beyond ourselves, to others; especially the needy, often neglected, and unloved pilgrims traveling life's winding pathways.

One of my greatest blessings came as a result of cleaning out my wardrobe of unnecessary clothing last Christmas. My friend Rick Van Hoose was preparing to return to Haiti after the holidays to help feed the starving masses near

Port-au-Prince where an orphanage sponsored by the Bible Center Church from my hometown of Evansville, Indiana, is located.

Rick shared with me shortly before leaving how most of the ministers lacked sufficient clothing and shoes to travel into the bush country to tell the natives about Jesus. This prompted me to sort out and send all my non-essential wardrobe with Rick, to be distributed among my black Christian fellow laborers in that primitive impoverished land. I received immense gratification through knowing I had played a minor role in touching these natives who I would never see or know in this world, by helping clothe God's messengers.

The Bible tells us we will receive a hundred-fold return in this present world for reaching out to share our resources. Put the pencil to this formula and see how many multiple blessings await the faithful. Touch one life a month and expect to be touched at an annual increment of twelve hundred (1200) times in return. We are not to reach out to others for the selfish purpose of being touched in return; however, when saints embrace others out of love, they can expect the handsome one hundredfold in return blessings. Isn't God good?

I believe Jesus would tell us, if He were here in the flesh, that He had never seen His touchers forsaken or their seed unloved. But a word of caution; touching can be contageous! If you take the chance to become a godly extrovert anointed with love, what you possess will be so attractive that others will covet this unique quality. This can cause a chain reaction, or domino effect. The domino effect has been popularized by contestants attempting to break the world's record by seeing how many thousands of dominoes can be downed by touching just one domino. The touched domino causes the remaining dominoes to be toppled by a chain reaction, igniting beautiful spectrums as the elaborately designed patterns are swept from their upright positions.

Are you following my trend of thought? If each born again child of the Most High would reach out and touch just one life, which in return would touch one life, the laws of touching would take effect, revolutionizing the world. Satan's engulfing clutch would be broken, setting its captives free. Ardent warfare would not be necessary. Instead, the all-powerful tender touch of believers touching unbelievers, stimulating them to become believers who in return touch other unbelievers, thus causing an endless chain of soul winners and instruments of encouragement. Victory could be won by the hand and heart which could never have been won by sword and shield.

Whenever there is a sincere desire to help others, God will see that the doors for service are opened. The results will be far beyond one's

greatest expectations. Again, it is utterly impossible to outtouch a God who set the laws of
touching in motion. I have tested these principles repeatedly and have yet to see them fail!
Let me share one account whereby a burning
desire to reach life's down and outers was
multiplied to encompass people from all life's
strata.

My last book, *You Are Somebody Special,*
started out as a minibook to be distributed free
of charge to men and women in the penal systems. As I opened my heart to reach these oft-
forgotten lonely individuals, God opened the
windows of Glory to pour out a fresh anointing.
As I viewed the vast need for all mankind to be
told they are special, nineteen chapters came
forth instead of one.

As the law of touching was put into effect with
the writing of the book, God threw in an extra
bonus; four chapters were edited to form a minibook entitled *Failures, Hookers, Prisoners Too,
You Are Somebody Special.* My desire to reach
life's suffering pilgrims of misunderstanding and
rejection was soon realized. The use and dis-
tribution of these minibooks was accepted by
several major institutional outreaches, including
Vicki Jamison's New Day House for Women, the
International Prison Ministry Association and
the PTL Club.

PTL just recently called to inform me that the
minibooks are now being included in the sal-

vation packet which is sent as part of their follow-up ministry to those who call or write, stating their decision to follow Jesus. Through God's laws of touching, a booklet born out of a desire to reach thousands is now reaching millions! God's Word cannot fail. If you do not agree, I challenge you to apply His principles to your own situation. If the motives of your heart are true, you will be astounded at the outcome.

6

SINCERE TOUCHING

My father was a true genius with an IQ to substantiate the fact. Unfortunately, it was not until I was twenty-seven that I began to realize just how smart he was. I'm not exactly sure what happened, but the year before I turned twenty-eight, Dad suddenly got smart!

Dad had many wise sayings. One adage I recall in particular was, "Dogs and babies are good judges of character." I have found this pearl of homespun truth to be quite valuable. Just stop for a moment and review the people you know. Now focus on those whom you know to be of less than stable character, or who have at least aroused suspicion. Have any of these people, to your knowledge, ever had counteraction with either dogs or babies — or perhaps both? While I am not claiming this test to be 100 percent valid, from my own observation I have found that it usually works.

Babies, as well as dogs, seem to have a built-in detection device that enables them to sense whether a person is genuine or counterfeit. Neither seems to be easily deceived by the false wooings of an insincere smile. As Jesus pointed out in Matthew 23, it is possible for one to look beautiful on the outside while harboring corruption on the inside. Or in the words of an old Negro spiritual, "Everybody talkin' 'bout Heav'n ain't a-goin' there."

God does not look upon the outward appearance, but on the heart, or inner man. It is not possible to fool our omniscient Creator who knows everything about everyone.

While dealing with the various aspects of touching, my purpose is not to stimulate people to touch as a mere form or ritual, but out of sincerity. But before genuineness can exist, honesty must be exhibited. Something as wonderful as the human touch can prove to be futile if not done out of empathy and concern.

People aren't stupid. As Abraham Lincoln once stated, "You may fool all the people some of the time; you can even fool some of the people all the time, but you can't fool all of the people all the time." Even though fakes may succeed in deceiving many, sooner or later they will be exposed.

Each of us is unique! Because of our different personalities, we display our emotions in different ways. Some of us are quiet and soft-spoken while others are outgoing and more boisterous. In psychology, we call the inhibited personalities *introverted,* while the exhibited personalities are labeled as *extroverted.*

Either personality can be effectively mobilized to reach out to others. Love is the key which opens doors for both the quiet and the not-so-quiet. Trying to change our basic personality would throw us into an altered state or unrealistic realm. Not everyone is cut out to be the life of the party. Some are destined to be part of the crowd, due to their psychological make-up.

While many of us are accustomed to using physical contact as an expression of inner warmth, there are others who feel awkward or uncomfortable with such closeness. By *natural personality,* I am not a toucher. However, by *spiritual motivation,* I am a warm person who comfortably touches and embraces others.

Have you just gotten confused by apparent double talk? Let me explain. Under the Holy Spirit's unction, we believers in Jesus can put on the mind of Christ (I Corinthians 2:16). We no longer do what we are used to doing, but we do what is pleasing to our brother, Jesus. There is a special love which engulfs mortal man with a supernatural capacity to care for others. Under this anointing, non-touchers become sincere touchers, motivated by a supernatural brotherly compassion.

Authors freqently share incidents which place them in an unfavorable light in order to help others learn from their mistakes. I must now do so, with great regrets for my attitude toward another individual. Although these feelings have been put under the blood of Jesus, it is still embarrassing to think how unchristian my attitudes were.

A couple of years ago, a man I knew was in critical condition in the Intensive Care Unit of a local hospital. This man and I were by no means friends. In fact, I felt he had offended me more than anyone I'd ever known. Because of this, I harbored deep resentment toward him. Or to put it more bluntly, I *detested* the man!

Knowing resentment had no place in my life as a Christian, I had prayed much about the situation. But in spite of my prayers, these feelings kept creeping back, attaching themselves to my spirit like barnacles. Perhaps I was asking God to alleviate a *condition* when He wanted to heal the *cause*, which was my spiritual thorn of hostility toward this person.

Perhaps it was because of this intense inward struggle that God chose me, of all people, to minister to this man. God often uses His servants to minister to those whom they formerly rejected, despised or avoided. Great men like Peter and Paul had to overcome personal hang-ups in order to follow God's mandates. Peter was directed to accept Gentiles whom he thought were unclean, while Paul became a convert to Christianity which he formerly persecuted unto death (Acts 10). God gave both of these men a sincere love for those whom they formerly hated. Knowing that I wasn't the only person with personal prejudices helped soothe my mind during this spiritual struggle.

I received a call late one evening informing me that this man's condition had worsened and he was bleeding profusely. His life could not be sustained for long due to the rapid loss of blood.

While I was still on the phone, God spoke to me that I should go immediately to the hospital and pray for this man. Surprisingly, I had no desire to argue with God or debate the issue. Instead, a sense of urgency was compelling me to rush to the man's bedside.

It's difficult to put into words what happened as I arrived at the hospital that night. As I entered the Intensive Care Unit I was suddenly bathed in a fragrance of supernatural love.

Words came with great difficulty, but I asked to pray with the pale, waning, soon-to-be corpse which lay before me. Dying people tend to let bygones be bygones, and desperate people are willing to bury the past and seek help from wherever possible — even from one's enemies.

I took the man's hand, somewhat awkwardly at first, and began petitioning heaven for physical healing. As I prayed, it was as though a floodgate had opened to allow rivers of love water to gush forth from my innermost being. Love was now radiating from a source where not even a spark had before existed!

I finished praying and stepped back to witness two miracles. The profuse bleeding had stopped immediately; and inside me, all the pent-up resentment had dissolved. One life received physical healing, while the other received spiritual restoration. And in case you are wondering, both healings were permanent!

God wants every man, woman, boy and girl to develop a sincere longing to uplift those around them. However, God does not promote false flattery. A compliment must come from the heart for the sole purpose of encouragement. "Buttering up" or "politicking" just to get ahead or to make points with someone is not only a form of dishonesty, it is a sin. Truth alone is pure, while shaded truths are impure, being tarnished by improper motives. The Christian's actions must be completely transparent, leaving no question as to one's intent.

Beloved, learn to like yourself so that you can like others. The basic reason why people distrust and dislike others is because of their own poor self-images. It is difficult, if not impossible, to reach out and care for others when we cannot stand ourselves. Jesus desires our inner healing so that we can become an extension of His love to others. Without it, we are like the blind leading the blind, causing both to fall into the ditch. When we feel good about ourselves, we become mobile "conductors" transmitting to others the power of God's touch.

No doubt the most simple, yet vivid discourse on touching is Jesus' own words as recorded in Matthew 25:31-46:

"When the Son of man shall come in his glory, and all the holy angels with him, then shall he sit upon the throne of his glory: And before him shall be gathered all nations: and he shall separate them one from another, as a shepherd divideth his sheep from the goats: And he shall set the sheep on his right hand, but the goats on the left.

"Then shall the King say unto them on his right hand, Come ye blessed of my Father, inherit the kingdom prepared for you from the foundation of the world: For I was hungered, and ye gave me meat: I was thirsty, and ye gave me drink: I was a stranger, and ye took me in: Naked, and ye clothed me: I was sick, and ye visited me: I was in prison, and ye came unto me.

"Then shall the righteous answer him, saying, Lord, when saw we thee an hungered and fed thee? or thirsty, and gave thee drink? When saw we thee a stranger, and took thee in? or naked, and clothed thee? Or when saw we thee sick, or in prison, and came unto thee?

"And the King shall answer and say unto them, Verily I say unto you, Inasmuch as ye have done it unto one of the least of these my brethern, ye have done it unto me. Then shall he say also unto them on the left hand, Depart from me, ye cursed, into everlasting fire, prepared for the devil and his angels: For I was an hungered and ye gave me no meat: I was thirsty, and ye gave me no drink: I was a stranger, and ye took me not in: naked, and ye clothed me not: sick, and in prison, and ye visited me not.

"Then shall they also answer him saying, Lord, when saw we thee an hungered, or athirst, or a stranger, or naked, or sick, or in prison, and did not minister unto thee? Then shall he answer them, saying, Verily I say unto you, Inasmuch as ye did it not to one of the least of these, ye did it not to me. And these shall go away into everlasting punishment: but the righteous into life eternal."

This is the type of touching that separates the sincere person from the phony. True, little recognition is received from man when we feed the hungry, lodge strangers, clothe the naked, care for the sick or visit the prisoner. God, however, takes notes of kind actions done in His name. These deeds are the direct channel by which we touch others with God's love. And eternal rewards are reserved only for those things done with sincerity from contrite hearts.

7

TOUCHING BY LISTENING

If you will observe people, you will note that those who are most popular are the ones who are good listeners. Their secret to social success lies within their ability to touch others by allowing them to be heard. People like to be heard when expressing ideas and opinions, or when relating personal experiences. Ideas are cherished by everyone and most people enjoy sharing their tidbits of wisdom with others.

By allowing others the privilege of complete expression, we in essence allow them to touch us. Failure to hear people out banishes their opportunity for expression of what they so desperately need to share and barricades avenues of communication with walls of rejection. For example, interrupting someone in the middle of their story conveys to that person that what they are sharing is of little importance, trivial or boring. Interruptions serve to emotionally push backward the one attempting to express their inner feelings. A failure to show an interest in what someone is saying is not just rejection of an idea, but of the person relating the idea.

A poor listener seems to be beyond the reach of the communicative touch of others. Attentive listening is the highest form of flattery, while ignoring what another person is saying comes across as a put down.

In the field of counseling, the primary principle to being effectual is the ability to listen. It is essential to meeting people's needs. It is virtually impossible to help a client unless the client is allowed to be heard. A counselor must allow himself to be touched in order to earn another's trust, otherwise his own opinions will have no value.

This same principle carries over into every field of human service. Successful teachers listen to the concerns of their students, thus enhancing the learning process. Effectual pastors listen attentively to the needs of their parishioners in order to minister to their needs. Top-level sales people gain the confidence of customers by first listening to their needs, then showing how their products can meet these particular needs. Politicians seeking their reelection listen carefully to the heartbeat of their constituency, thus expressing their ability to be touched by those whom they represent. All those who depend on others for their livelihood well realize the importance of making themselves accessible to their constituents, listening carefully to those who provide their support.

Bob Crisp spearheaded the Winners' Circle Division of the Amway Corporation into one of the greatest business success stories of our time. He did so by teaching his salespeople not to concentrate on selling Amway products, but to sell prospective customers on their self-worth and dignity as individuals. Encouraging others to discover their unlimited potential and listening to their needs, dreams and desires allowed Crisp and his representatives to enlist a mammoth sales force. These direct sales experts were catapulted to the top in their profession because they allowed themselves to be touched by being good listeners. This unique characteristic was coupled with the encouragement of individuals to be sold on *themselves*. The end result was an army of self-confident, attentive-listening, tactile business representatives who would increase annual sales from *nothing* to over one hundred and twenty million dollars in just six short years, which seems unrealistic. This fairy tale ending was achieved because the tools of listening and touching were put into practice.

Dale Carnegie courses have helped millions

to *win friends and influence people* by teaching the principles of good listening and self-encouragement by establishing good rapport. The ability to remember a person's name makes that person feel important. So does the attentiveness of one who will listen and focus the attention on the other party rather than on the familiar territory of self-achievements. Human nature dictates desires to be heard which often go unfulfilled because of the inability of people to listen.

Would you like to win someone's friendship and admiration? Then ask that person about himself and explore his interests. Let him do the talking while you do the listening, and you will touch that person and make him feel like a "somebody." Nothing is more dear to the heart of a person than *he* or *she* or *his* or *her own*.

Inquiring about a person's interests makes that individual feel important and special. Every stable person alive needs to feel special, and what can boost a person's feeling of self-worth more than exploratory conversation? This doesn't mean you must deny yourself the privilege of sharing your own thoughts. But follow the advice of the apostle Paul to "be kindly affectioned to one another in brotherly love; in honor preferring one another" (Romans 12:10).

Television talk shows have enjoyed increasing popularity over the years. This is because of the good rapport the hosts of these programs are able to create between audience and guests as popular topics are being discussed. The key element of a successful talk show is a perceptive host — one who can direct the conversation while allowing his guest to do the talking.

I consider Jim Bakker to be one of the best talk show hosts in the world because of the way he permits his guests to express what is on their minds. Jim is by nature an attentive listener, and by making his guest feel comfortable, he makes the audience comfortable with the guest.

Another outstanding quality contributing to Jim's success as an interviewer is his refusal to slight lesser-known guests for well-known celebrities. I have had the privilege of sharing the PTL set with both Pat Boone and singer Connie Smith Haines; but I was in no way made to feel less important simply because I am not as well-known. From personal experience, I can assure you this makes a lesser-known person feel great!

Having done talk shows from coast to coast, I can say nothing has been more frustrating than being interrupted mid-sentence by a host who preferred to monopolize the conversation. Without exception, the hosts most popular with their viewers are the ones who can maintain control and yet direct the conversation's focus toward the guest for the benefit of the audience.

Two other men whom I consider real pros in

the field of television are Steve and Lester Sumrall. Being a frequent guest on their one-hour show stimulates advance preparation, because I know at least three-quarters of that time will be spent answering questions. These men affect millions of viewers by being good listeners and thus successful television hosts.

Also among my favorite hosts are Tommy Barnett of *Praise* in Phoenix, Jerry Barnard who often hosts *Praise The Lord* in Los Angeles and David Mainse of Toronto's *100 Huntley Street.*

Tommy Barnett summed up the principle of listening once when we were talking about helping people. "Our job," he said, "is to listen to the hurts and needs of others and then help heal those hurts." This is touching by *listening* in the truest sense. People cannot be helped until they are reached, and their needs remain unknown until they are heard.

A good listener will discover that most problems can be solved merely by listening to what a person has to say. Usually the person talking will answer his or her questions if they are given a chance to "talk out" what is troubling them. What they need is a sounding board in the form of a person who cares and expresses concern for their well-being.

Hundreds of people who write to columnist Ann Landers end their letters by saying, "thank you for listening." By listening, one is silently saying to that person, "I care about you. I am interested in your ideas and your concerns. You are important and I am willing to be touched by you."

A good conversationalist will shift the focus of the conversation to others rather than self in order to touch their interest levels. Efrem Zimbalist, Jr. is one of Hollywood's most polished conversationalists. Efrem has a wit able to spellbind most, and could focus upon his own career and never bore his listeners. Yet this precious, sensitive man constantly shifts the conversation to others, because he is more interested in others than in himself. Working and fellowshipping with Efrem has given me a deep appreciation and respect for him. Never have I met a man more genuinely interested in people and less desirous of promoting self. This unique attribute enables him to effectively touch all those with whom he comes in contact.

If you are sincere in your love for people you will find them interesting and worthy of being heard. Once you cultivate good listening habits you will find people being attracted to you. It is easy to love a good listener because he demonstrates love by being willing to be touched. His responsiveness makes him accessible to the extended souls of all who touch him.

Popularity could well be said to rest upon three basic principles. *Sit* up, *listen* up, and *shut* up until the other person is finished talking! By sitting up, we show our respect and demonstrate an interest in the conversation of others. By listening up, we allow others to express their ideas freely while we demonstrate a genuine concern for what they have to say. By shutting up until the other person has finished talking, we place that person in a position of primary importance by putting his ideas ahead of our own. Every person feels his ideas are the best, so why not permit him this privilege? After all, we already *know* our own ideas are *really* the best, so why not let others feel the same way?

Touching by listening may at first be rather difficult; especially for the more talkative person. However, once mastered, it can prove to be the most rewarding endeavor ever undertaken. Making others feel important will bring more satisfaction than words can describe. If you want to become a full-fledged toucher, start by becoming a good listener. It can be fun! Why don't you try it and see what happens?

8

TOUCHING LIFE'S UNTOUCHABLES

In biblical times, when a person was found to have the dreaded disease of leprosy, he was labeled "unclean" or "untouchable" and was made to live outside the camp until such time that he could appear before the priest and be declared clean. He then had to bring a proper sacrifice and go through certain rituals of cleansing before being admitted again to the mainstream of society.

Today the cruelty of life has created many types of lepers who most prefer to ignore; certainly not to touch. We have a tendency to make our own "priestly" judgment and ostracize such ones from our company until their lives, in our opinion, are clean. How tragic! Who are we to call one of God's creation untouchable, unclean or insufficient? Every person is a masterpiece designed by God and created in His image. God never ordained man to reach some of the world, but to reach the entire world without regard to another's worthiness or unworthiness. This mission weighed so heavily upon the heart of God that he delayed the return of Jesus until the gospel of the kingdom has been preached to all the world (Matthew 24:14).

Every person on the face of the earth is a ready candidate for touching. Just as every man, woman, boy and girl deserves to hear the salvation story, so does each have the God-given mandated right to be reached by Jesus' "touch brigade:" caring Christians. Entrusted to believers is the awesome responsibility of reaching fallen lives. Jesus did not "suggest" that His children reach sinners; He commanded," ... You are to go into the world and preach the Good News to EVERYONE, EVERYWHERE" (Mark 16:15 TLB).

If you attempt to follow this heavenly directive, don't necessarily expect to be popular with everyone, or to be understood by many so-called Christians. Our Lord Himself was ridiculed by the religious leaders of His day because of His association with tax collectors and other riffraff. It seems preposterous that one who claimed to be the Messiah would surround himself with thieves, liars, prostitutes, beggars, fornicators, and all sorts of undesirables. The answer was simple as Jesus explained, "I am not come to call the righteous, but sinners to repentence" (Matthew 9:13). Jesus came to reach those who longed for His gentle touch. This completely boggled the minds of the self-righteous religious leaders, so steeped in tradition that they were fanatics in their "untouchiveness."

The Pharisees and Sadducees would rather have been found dead than touching a corpse, gentile, social outcast, or anything else considered ritually unclean. Jesus blasted those who hid behind a facade of godliness while showing unconcern for their fellow man (Matthew 23). These men had vainly attempted to adhere to the letter of the laws of Moses, while ignoring the intent of the laws of God. They had elected to follow the empty rites of abstaining from things defiled, while blinded to the fact that God had combined all Ten Commandments into two which dealt directly with touching. The first commandment, as taught by Jesus, was for man to touch God by loving Him with all his mind, body and soul. The second was for man to touch his neighbor by loving him as himself (Matthew 22:37-38).

Our love for God is demonstrated by touching the lives of those whose paths we cross. "If a man say, I love God, and hateth his brother, he is a liar: for he that loveth not his brother whom he hath seen, how can he love God whom he hath not seen?" (I John 4:20). Beloved, it is time for Christians to put up or shut up. James also admonished, "Even so faith, if it hath not works, is dead, being alone. Yea, a man may say, Thou hast faith, and I have works: show me thy faith without thy works, and I will show thee my faith by my works. Thou believest that there is one God; thou does well: the devils also believe, and tremble. But wilt thou know, O vain man, that faith without works is dead?" (James 2:17-20).

Perhaps the most graphic account of touching ever written is the story of the Good Samaritan (Luke 10). The hero of the story is a Samaritan, a man despised by the Jewish religious hierarchy for ethnic reasons; yet he put these men to shame in his obedience to God. This half-breed was regarded as a bastard by those adhering to the letter of the law, yet he had a spirit of compassion which compelled him to befriend the beaten, wounded stranger who had been left for dead in the dust by the roadside. The Good Samaritan touched the heart of God by reaching out to a stranger who had been totally ignored by the very ones expected to respond to a state of distress.

Jesus used this story to illustrate that everyone is our neighbor, and we are not afforded the luxury of "selective" touching. True Christians cannot say, "I'll touch the clean, but not the dirty; the moral, but not the immoral; the sober,

but not the drunk; the wealthy, but not the poor;
the free, but not the imprisoned; the majority,
but not the minority; the sane, but not the mad;
and people of my persuasion, but not those of
different beliefs."

Biases and prejudices can hamstring one's
ability, thereby limiting their capacity to effectually reach those who are different. The
apostle Peter had such a hang-up until God
opened his spiritual eyes through a vision. Come
with me to the town of Joppa, to the lakeside
home of Simon the Tanner. Peter has come here
for a time of rest and spiritual refreshing. As we
observe from the rooftop, Peter, the Rock, ap-
proaches to meditate and communicate quietly with "HIS" God. Of course, Peter's private
God was Jewish; after all, "How could the Creator of the universe have anything to do with
those pagan Gentiles?" Peter had reasoned.
Can't you just visualize Jesus sitting at the right
hand of the Father, grieving at what He sees in
the heart of this man who had been part of the
"inner circle"; a man who had been an eye-
witness for more than three years as He healed
Gentiles and forgave them their sins?

I can picture, in my mind, what transpired
that day. A glistening tear sparkles in the radi-
ance of heaven's bright splendor, as it runs
down Jesus' cheek. Quickly He entreats, "Holy
Spirit, show Peter that the God of Abraham,
Isaac and Jacob is also the God of the Gentiles.

Tell him that they too were created in OUR image and should not be regarded as inferior or untouchable. And tell Peter that he and the other apostles must reach out and touch the despised pagans, because I died for them as much as for the household of Israel."

What rejoicing must have echoed through the portals of Glory that day! The angels, who rejoice when salvation comes to one lost sinner, gazed into the future and beheld millions of souls, once denied the message of grace, now being accepted as divine citizens of the household of faith. It had not been a matter of the Gentile world hating God, but rather that they could not possibly believe in a God of whom they had not heard until being reached with the good news.

Meanwhile, back on the rooftop, the gentle breeze rustles through the swaying trees, casting silhouettes which fall in ghostly shadows across the roof. It is a lazy day as Peter reclines in the warmth of his pinnacle abode. Silence abounds except for the occasional yelp of a neighbor's dog and the gentle lapping of waves along the shore. Basking in the peaceful quietude of this setting, Peter begins to nod before falling into a deep trance. Wait! Who is that with Peter? He is no longer alone on the roof as visitation is made by God's Spirit.

Let's read this beautiful account together from God's Word: "And saw heaven opened, and a certain vessel descending unto him, as it had been a great sheet knit at the four corners, and let down to the earth: Wherein were all manner of fourfooted beasts of the earth, and wild beasts, and creeping things, and fowls of the air. And there came a voice to him, Rise, Peter; kill, and eat. But Peter said, Not so, Lord; for I have never eaten any thing that is common or unclean. And the voice spake unto him again the second time, What God hath cleansed, that call not thou common. This was done thrice: and the vessel was received up again into heaven" (Acts 10:11-16).

9

A POINT OF CONTACT

As a child, I well remember the Oral Roberts tent services which were televised. This was before the days of color TV and the vast array of Christian programming seen today. Sunday afternoon always found our family seated in front of a little black and white TV set.

In every service, Oral would first preach a solid scriptural message — then the healing lines would form. "I am going to reach out and place my hands on you as a point of contact," Oral would say. "I cannot heal you — only God can do that." Then seated on a stool, he would face the person with the need and place his hands on either side of their head as he prayed.

Quite often he would encourage those listening at home to reach out and touch their radio or television as a *point of contact.* Even with my regimented Baptist background, I never doubted people being divinely healed in Oral's meetings — however, his grabbing people's heads when he prayed caused me to shudder in dismay. I was equally confused by his reference to a point of contact. So even though I knew his teaching was scriptural, his methods turned me off.

It is quite easy to be against those things which are foreign, fearful or threatening to our doctrinal positions, and like most prejudiced people, I did not accept that which I could not understand. In fact, some twenty years passed before I would come to understand Oral's *point of contact.* Meanwhile, I would engage in many struggles, with unnecessary victories being lost because I had not allowed myself to be receptive to the touch as a point of contact.

Until 1975, my spiritual life was like an amusement park rollercoaster, clanking up one hill only to plunge down into a valley. Then so much happened that year and the one which followed that it took a whole book to tell just a part of what occurred. This story is told in more detail in *A Glimpse Of Perfection,* but since my purpose here is to set the stage for the touching discovery, I will only touch briefly on what happened.

In March of 1975, our two-year-old son Robbie was admitted to the hospital with a deadly virus which was sweeping the nation, killing many children and elderly people. Panic set in when his temperature shot up to 104 degrees with negative responses to all medication. After four days of severe vomiting and diarrhea, Robbie lost one-third of his total weight and became so weak that we were desperately reaching out to touch the hem of the healer's garment. My wife Colene and I both promised God our lives unreservedly if He would spare Robbie's life.

Two days later a dramatic turnaround took place and we took our son home, weak but well. Exactly six months later, Jesus appeared to me, completely revolutionizing my life. At His appearance, my dormant faith was released to operate with the accompanying power of the Holy Spirit. I was about to become a toucher!

As I looked into the piercing, loving eyes of Jesus, a new dimension of His love was revealed. An instantaneous empathy for others was established within a baptism of divine love. Jesus' supernatural personage was overwhelming. It was like a giant tidal wave of sweet perfume came billowing across my soul.

In the holy presence of Jesus, I felt naked, ashamed, abased and fearful, due to my carnal imperfections. Yet the attracting warmth and love of the Master drew me to Him like a magnetic force. I felt an inner warmth which was beyond description— like warm oil being poured over my body. Yet at the same time, chills ran up and down my spine. I thought, how can such imperfection stand in the presence of complete perfection and survive? The answer was simple. In my mind, I could visualize Jesus with arms outstretched — reaching out to *others* through *me*!

During the next six days, I was able to see future events through dreams and discovered the significance of touch as a point of contact.

The same night that Jesus appeared, I had a dream in which Colene was lying in a hospital room. The attending physician was saying, "I'm sorry — it's terminal." Frantic at the prognosis, I muttered, "Unless God intercedes."

Less than thirty-six hours later, her appendix had begun to burst, spreading infection through her body. Emergency surgery was undertaken, but by then Colene was a mighty sick lady and the doctors prescribed a seven-day stay in the hospital for recuperation.

Three days after surgery, two friends came to visit Colene in the hospital. Before they left, we all three reached out and touched her as we asked Jesus to give her a speedy recovery. Then around 6:30 the following evening, the doctor unexpectedly dropped in and dismissed her from the hospital. Still, a full comprehension of the touch as a point of contact was not crystalized.

A few weeks later the picture became clear as

a mountain stream.

It happened when I was invited to speak at a Christian businessmen's meeting. Having never spoken before such a group before, I was dismayed when they told me I was to "minister" and not preach. My frustration mounted when I misplaced my notes just minutes before being handed the microphone. (God must have chuckled when He hid those notes in that old Thompson Chain-Reference Bible!)

As I stood before the group, in my own eyes everything seemed awkward and unbalanced. This was because God had suddenly displaced me at the controls — something which had never before happened in my life. I felt so self-conscious and humiliated by my blundering presentation that all I wanted to do was quickly finish what I had to say, give a quick altar call and get out while the gettin' was good!

I suppose every speaker feels like he has done poorly at one time or another, but that night I had the feeling I had bombed out royally! After all, I had never experienced walking in the Spirit and the sensation was strange — almost spooky!

When I gave the altar call, I was actually surprised when people responded. As folks came forward for prayer, the Spirit compelled me to touch each one while praying for their needs.

I stood dumbfounded as miracles began to happen right before my eyes! Suddenly it was as if a time machine had transposed the scene to the apostolic days when miracles were being manifested before the eyes of the multitudes. These unexpected supernatural happenings were not confined to some remote corner of the room, but were taking place up front for all the people to see. I learned that God, unlike so many of His servants, is not the least bit bashful.

Standing with the aid of a cane near the back of the room was a man severly crippled from multiple sclerosis. As I walked toward him, he began a slow, torturous journey in my direction, twisting his pain-wretched body forward and dragging one leg in order to move.

"I want to be healed," he said pitifully as we met, "Oh God! I want to be healed!" The Spirit directed me to pray for his healing. I timidly extended my right hand to touch his head as I prayed — but much to my dismay, nothing happened. Then the tender voice of God said, "Face him and place your hands on the back of his neck as you pray for him."

I felt embarrassed by the apparent failure of my first prayer, so it was easy for me to be obedient and pray again. As I stood in front of the crippled form with my hands reaching for his neck, visions of Oral Roberts' tent services flashed through my mind. Years of prejudice suddenly vanished and I thought, *If only Oral Roberts could see me now!*

While I prayed, it felt as though thousands of volts of electrical current were being discharged through my body to the crippled man. "I feel like my body is on fire!" he shouted. "I'm healed! I'm healed" The fact that he was standing for the first time in years without aid of a cane was in itself evidence that he had been healed, but he couldn't wait to try walking.

He started out at a slow, careful pace — praising God with every step. Then he picked up momentum as he scurried across the room, and before he was finished, he had jubilantly run several laps around the room!

All the while, I stood watching — amazed at the tool of *touch* as a point of contact. I realized that neither I, nor Oral Roberts, nor anyone else, had within ourselves the power to heal. Yet, because I had been obedient in touching, God had perfomed a mighty miracle!

This phenomenon is referred to in the New Testament as the "laying on of hands." It serves as a point of contact through which supernatural spiritual manifestations are transmitted.

The laying on of hands was first instituted by Jesus and His disciples and the practice was continued by the first century church. This ordinance was practiced in ordination services of minsters and deacons as well as in services where the sick and afflicted were being prayed for to receive divine healing or deliverance.

The apostle James wrote, "Is any among you

afflicted? let him pray. Is any merry? let him sing psalms. Is any sick among you? let him call for the elders of the church; and let them pray over him, anointing him with oil in the name of the Lord: And the prayer of faith shall save the sick, and the Lord shall raise him up; and if he have committed any sins, they shall be forgiven him" (James 5:13-15).

God's healing, both physical and spiritual, still flows today through the touch as a point of contact. Our God is the same yesterday, today, tomorrow and forever.

Trying to figure out why some receive healing while others go unhealed, even after prayer warriors had laid hands on them, caused me a great deal of confusion until God revealed to me one simple law concerning healing. I call it *the law of positive and negative faith.*

It works this way: If I have positive faith and lay hands on you, praying for your healing, the results will be determined by your faith. If your faith is absolutely *positive,* you will experience healing. If your faith is *neutral,* healing may still be manifested. But if your faith is *negative,* healing will not be experienced. The touch is a unifying factor operable when two or more become one in the spirit of supernatural expectations known as faith.

A few years ago, I heard one well-known radio minister say, "The only thing you will get from the laying on of hands is germs." I was saddened

by this statement to which I take complete exception. When one touches a fellow believer in
a prayerful attitude, the human touch serves as a
unifying factor capable of stimulating mutual
confidence in both parties. Love is transmitted
through the touch which is also a channel through
which blessings can flow. Psychologically, touching renders a boost to an individual by showing
that person he is important enough to be reached out to. Spiritually, touching is the connective
element through which the Spirit of God gener-
ates redemptive regeneration and restoration of
the spirit, mind and body.

In my denomination, Southern Baptist — as
well as in many others — laying on of hands is
practiced at ordination services for deacons and
ministers. This practice signifies that the person
being ordained has been approved for service
by the governing body of ordained officials who
are asking God's blessing on the person being
set forth.

The ritual is climaxed by all of the ordained
officials laying hands on the candidate for a
prayer of exhortation in which God's blessings
are sought. Unity is expressed by the body
which touches the newly-ordained person. Also,
a point of contact is established between the
group and heaven on behalf of the church and
the one it is sanctioning to represent the Kingdom to the world. Brotherly love and encouragement are transmitted in a most sacred way

through the presbytery's touch as a point of contact visible to both man and God. As Paul admonished Timothy, "Neglect not the gift that is in thee, which was given thee by prophecy, with the laying on of the hands of the presbytery" (I Timothy 4:14). So should God's servants be encouraged today!

In the last twenty-five years, the practice of touching as a point of contact has been revitalized internationally. When used as a launching pad for the release of dormant, inactive faith, the touch is spiritually beneficial and rejuvenating. The human touch has a therapeutic value which can be obtained in no other manner.

We can see faith being propagated through the touch as the stronger person touches the weaker. As the one's faith is released, the other's faith begins to develop. When we see these Christian attributes passed on through a spiritual chain reaction, we can see God's children grow and become what He intended for them to be.

Beloved, I would encourage the laying on of hands only if God so directs. There are multiple gifts of the Spirit as I Corinthians, chapters twelve and fourteen teach. Some servants, in order to see results, are directed to lay hands on those for whom they are praying; others are not. Each has to operate within the framework of his spiritual calling.

Paul warned Timothy to "lay hands suddenly

on no man" (I Timothy 5:22). This sacred act is not to be practiced impulsively, but only under the unction of the Holy Spirit. A gift designed to bring encouragement would only bring reproach if hands were laid on by human impulses rather than by divine guidance, yet man can never fail when acting under God's mandates. When He directs His servants to lay hands on the needy, their needs WILL be met.

We must be obedient putting God *to the test*, but never putting Him *on the spot* by acting on our own volition or personal feelings. God has yet to fail or to permit one of His servants to fail when they are abiding in the center of His "perfect will."

Although we are to shun the reckless laying on of hands, we are to boldly do so when being wooed by the gentle voice of the Spirit. Our touch could be the vital contact needed to complete the connection between earth and heaven — the determining factor between defeat or victory for some poor soul needing help. Since Jesus went to sit at the right hand of the Father, millions of miracles have taken place through His extended touch by His body of believers.

We are the hands Jesus works through today to accomplish the work of the Kingdom, because man is the only creation counted worthy to serve as God's mouthpiece proclaiming salvation's story. Won't you reach out to others when God chooses *you* to be His point of contact? Jesus tells us through His God-breathed word, "Again I say unto you, That if two of you shall agree on earth as touching any thing that they shall ask, it shall be done for them of my Father which is in heaven" (Matthew 18:19). With such a powerful mandate, how can we as believers ignore the point of contact as a ministry?

Our touch has not been suggested, but *ordered* of the Lord! On the basis of this scripture, if you and I earnestly desire the Lord to intervene in our behalf, we must be willing to reach out with another believer to touch that which we seek. When we yield ourselves to the touch as a point of contact, God has promised to honor our petitions by removing obstacles and altering impossible circumstances.

Many sermons have been preached and countless books written regarding the tongue as a positive force. Unfortunately, the importance of the touch combined with positive confession has been ignored. I admonish you to employ the union of touch and prayer when petitioning God for divine intervention. If possible, lay hands on the situation; if not, touch the situation with your heart and mind asking God to complete the desired work. By establishing a point of contact when voicing our request to God, we are being specific as well as obedient.

If you will read closely, you will find that the

woman with the blood disease *touched* Jesus before she asked to be healed. She established a point of contact signifying her faith. Her touch resulted in healing powers being transmitted from Jesus before He even turned to ask who touched Him and what her need was. Let's read this passage together from Mark 5:25-34.

"And a certain woman, which had an issue of blood twelve years, And had suffered many things of many physicians, and had spent all that she had, and was nothing bettered, but rather grew worse, When she had heard of Jesus, came in the press behind, and touched his garment. For she said, If I touch but his clothes, I shall be whole. And straightway the fountain of her blood was dried up; and she felt in *her* body that she was healed of that plague. And Jesus, immediately knowing in himself that virtue had gone out of him, turned him about in the press, and said, *Who touched my clothes?* And his disciples said unto him, Thou seest the multitude thronging thee, and sayest thou, Who touched me? And he looked round about to see her that had done this thing. But the woman fearing and trembling, knowing what was done in her, came and fell down before him, and told him all the truth. And he said unto her, Daughter, thy faith hath made thee whole; go in peace, and be whole of thy plague."

Now that we've surveyed the importance of touching to establish contact points for release of faith, won't you begin to touch those things which need changing in your life, as well as in the lives of others?

10

A PAT ON THE BACK

Most people respond more favorably to a pat on the back than to a swift kick in the pants! Human nature craves positive reinforcement. Everyone desires praise from others; especially the ones they love and respect. The soul of mankind longs for loving strokes which can satisfy the need to be touched and encouraged.

If dogs respond to a pat on the head from their masters, how much more will humans respond to encouraging, uplifting, positive strokes from their fellow beings. No one in their right mind wants to be "put down!" Everyone wants to be lifted up and supported.

During a career spanning nearly two decades, I have worked with hundreds of clients who have negative self-concepts. Tragically, many have been dropouts from various churches whose programs failed to effectually touch them. Nearly all have had one common determining factor. They have not been touched physically, spiritually, or emotionally by the ones they long to have notice them. One little pat on the back by a parent, teacher, or pastor could perhaps have made the difference between their withdrawing into an inner shell or developing a positive self-image.

The most powerful force available is positive words. Words of encouragement can accomplish what nothing else can do. Successful coaches well realize the importance of positive reinforcement (praise) along with the encouraging physical touch.

Observe some of the outstanding coaches on television and see how often they pat their player on the back when the going gets rough. When the game is tied with the bases loaded and no outs in the World Series, watch the veteran coach as he strolls to the mound to encourage his pitcher. After a few words of advice, he says, "I know you can do it, you've got what it takes. Now strike him out with your superduper fast ball! The whole team's behind you — you're the best and those batters don't stand a chance against your speed!" Then the coach pats the pitcher on the back before leaving a super-charged hurler on the mound to rapidly retire the batters in order. Momentum has been increased by a pat on the back and the result is victory for the team.

Those who are not afraid to touch others can be described as *motivators*. Motivators are brave individuals who stimulate those with whom they associate to excel and become achievers. Achievement is seldom accomplished without the achiever having first been motivated to exert the necessary effort to accomplish the task.

Worthwhile victories do not come without hard work to overcome obstacles. No soldier is entitled to wear the combat ribbons unless he is a veteran of the war. Victory circles are reserved for those willing to endure the long grueling hours of training which helps them become winners. While often the ones responsible for motivating these winners receive far less recognition or reward, they are the *true* winners.

Don't you enjoy hearing a recognition speech when the recipient acknowledges the one who touched his life and inspired him to become a winner? Most gracious achievers willingly give credit to the ones responsible for directing them toward the finish line of success.

I enjoy watching award presentations on television because nearly all the recipients give homage to the ones who touched them and made their award possible. The ones behind the scene made it possible for their "touchees" to bask in the limelight of glory. True friends and genuine Christians are not concerned about who gets the glory, but that the job gets done.

A real friend is one who is willing to pat his colleague on the back even though it may catapult the friend ahead of himself. Several years ago, tenor singer Lee Robbins asked an unknown singer to perform during one of his concerts. Lee was willing to share the limelight because he had concern for the singer who was struggling to get his career started. The singer Lee touched was none other than Andrae Crouch, whose fame as one of the greatest Christian songwriters and vocalists is now international. Who knows what the result of a pat on the back can be?

Just this morning I ran into Bob, an old friend of mine on the police force. As we talked, I mentioned to him that I was writing a book about *touching.* My comment unlocked the door for him to begin pouring out some of the hurts and heartaches caused by two broken marriages.

The biggest hurt of all centered around Bob's teenage son who had recently returned to his father after living a number of years with his mother and her boyfriend.

"When my son first returned," Bob told me, "he withdrew every time I tried to put my arm around him to hug him or even touch him." He explained that the boy had become so accustomed to being pushed aside and verbally

browbeaten that he escaped by retreating into
a shell. The mother didn't believe in any physical
demonstration of affection toward her son, so
he felt rejected by her and became jealous and
fearful of her boyfriend.

"It has taken the boy several months," Bob
said, "but he has finally come out of his shell.
Now I can touch him and express love without
him shuddering. For months he would shudder,
grit his teeth and pull away if I so much as placed
my arm on his shoulder."

What a tragedy! This case is just one in a
million where a pat on the back could have
changed an entire lifestyle. Man's adamant nature often drives God's greatest creation, mankind, to push others down into the quagmires of
life. Unredeemed man has a natural inclination
to elevate himself at the expense of others by
walking over them or using them as rungs while
climbing the ladder of selfish success. Contrary
to a number of popular philosophies, unredeemed man is *not* basically good. Jesus, whom
I consider to be a behavioral expert, said, ". . .
there is none good but one, that is God" (Matthew 19:17). Man in his natural state is evil. His
depraved condition is altered to take on characteristics of goodness only through salvation.
The act of accepting Jesus as Savior qualifies an
individual to take on the mind of Christ, which
replaces natural imperfection with spiritual perfection.

The contrite gesture of a simple compliment can make the difference between failure and achievement. Multiple hosts have never excelled because they were never encouraged by someone who cared or had confidence in their performance. Human nature lends itself to exerting the additional effort needed to live up to the expectations of those whom we respect and admire. This is why we see tasks accomplished far beyond the imagination, when someone believes it can be done and encourages another to believe in himself enough to do the seemingly impossible.

Stimulating another person to believe in himself is the greatest gift mortal man can give another human being. No greater humanitarian service can be rendered than causing a child of the Most High to discover his potentialities or to reinforce his self-confidence.

Personal dignity as an individual who is unlike any other creation, can be increased by the encouraging words of a true friend.

The writer of Proverbs describes a friend as one who "... loveth at all times" (Proverbs 17:17). And for an eloquent picture of love, let's read the apostle Paul's letter to the church at Corinth.

"If I had the gift of being able to speak in other languages without learning them, and could speak in every language there is in all of heaven and earth, but didn't love others, I would only be making noise.

"If I had the gift of prophecy and knew all about what is going to happen in the future, knew everything about *everything*, but didn't love others, what good would it do? Even if I had the gift of faith so that I could speak to a mountain and make it move, I would still be worth nothing at all without love. If I gave everything I have to poor people, and if I were burned alive for preaching the Gospel but didn't love others, it would be of no value whatever.

"Love is patient and kind, never jealous or envious, never boastful or proud. Never haughty or selfish or rude. Loves does not demand its own way. It is not irritable or touchy. It does not hold grudges and will hardly even notice when others do it wrong.

"It is never glad about injustice, but rejoices whenever truth wins out.

"If you love someone you will be loyal to him no matter what the cost. You will always believe in him, always expect the best of him, and always stand your ground in defending him.

"All the special gifts and powers from God will some day come to an end, but love goes on forever. Some day prophecy, and speaking in unknown languages, and special knowledge... these gifts will disappear.

"Now we know so little, even with our special gifts, and the preaching of those most gifted and still so poor. But when we have been made per-

fect and complete, the need for these inade-
quate special gifts will come to an end, and they
will disappear.

"It's like this: When I was a child, I spoke and
thought and reasoned as a child does. But when
I became a man, my thoughts grew far beyond
those of my childhood, and now I have to put
away childish things.

"In the same way, we can see and understand
only a little about God now, as if we were
peering at His reflection in a poor mirror; but
someday we are going to see Him in His com-
pleteness, face to face. Now all that I know is
hazy and blurred, but then I will see everything
clearly, just as clearly as God sees into my heart
right now.

"There are three things that remain . . . faith,
hope, and love . . . and the greatest of these is
love" (I Corinthians 13 TLB).

Children especially need positive strokes from
adult authority figures whom they revere. This is
why children constantly watch the facial expressions of their parents or adult role models
when attempting difficult pursuits or trying new
challenging feats. Expressions showing confidence and approval of the adult are transmitted
into messages of encouragement to the youngster attempting to try out new fragile wings in
first flight.

When faced with unfamiliar surroundings or a
new challenge nothing reassures a frightened

child more than a message of confidence. This may be in the form of a wink, smile, a pat on the back, or a simple, "I know you can do it."

We all crave approval; we need to know that someone believes in us as a person and in our ability to perform. It is so much easier to believe in ourselves when others believe in us also. Confidence can be wonderfully contagious! For example, if I believe in you enough to build your self-confidence, you will generate enthusiasm toward others, causing them to believe in themselves and they in turn will inspire others. A chain reaction of self-confidence could be formed capable of circling the universe. Encouragement is a healing ointment to the person plagued by insecurity. Why should he settle for failure when success could just as easily be his reward? Personal encouragement stimulates enthusiasm and self-confidence which are essential to productivity.

Many businesses are learning that productivity levels increase when their employees are motivated by programs designed to reach them as individuals rather than regarding them as mere computer numbers. Seminars focusing on individual creativity are being used effectively in their work.

Big business has reaped the benefits of personal motivation in several ways. Not only has productivity increased, but absenteeism has sharply declined when employees have been involved in personal motivation seminars and group discovery sessions. Developing pride in one's work causes individuals to view services being performed as contributions rather than meaningless tasks for which compensation is received. Personal pride is a greater stimulus toward achievement than money alone.

In 1981, the Ford Motor Company incorporated the importance of personal pride in its national advertising. They produced television commercials which showed inspectors personally signing inspection papers on each automobile coming off the assembly line. The commercials were accented by the inspectors saying, "I take pride in what I do, and when I put my name on this baby, it comes back to me if you have trouble!"

The personal touch is needed in all facets of life from business to church. There is no suitable substitute for the element of one person reaching out to another with an extended helping hand. Impersonality breeds contempt but tactility stimulates love.

The depersonalization of individual dignity which the computer has spawned needs to be rectified by people touching people, restoring interpersonal contact. An emotionally hybrid generation will soon be created with monstrous characteristics unless mankind becomes willing to extend the vital ingredient of the complimentary human touch. Without this vital ingredient, the heart becomes a sterile being, void of love's tendering effect upon which all life depends.

Award-winning coaches, without exception, have learned to stimulate their players by patting team members on the back and building their personal confidence. Motivated players put forth the extra effort needed to become winners because they have learned to believe in themselves because their coach believes in them. Let's apply their example to our own lives. Let's begin seeing the people around us as "our team" which we are going to motivate by giving the players a *pat on the back!* We are going to encourage each member with positive motivational comments. *We* are going to demonstrate our love by speaking praise and by physical pats on the back! If we will dare to accept this challenge, the world will see an army incapable of accepting defeat. Look out world, here we come!

11

THE TENDER TOUCH

The word "tenderness" connotes qualities of softness, caring and delicacy of touch.

Unfortunately, many songs, plays and works of art have set forth the pseudo-hypothesis that being tender is synonymous with being a weakling. But compassion should never be mistaken for weakness. Strong men cry and strong women sometimes faint.

Have you seen athletes shed tears of joy or disappointment? Tears are a natural response which should not bring embarrassment or shame. As one songwriter puts it, *"tears are a language God understands."* Perhaps tears flush the inner man, ridding him of stockpiled frustrations.

It is a proven fact that crying brings emotional

satisfaction and relief. Since we associate tears with pain or disappointment, this may seem like a strange or alien statement. If so, let me ask you a few questions.

Have you ever wept at the movies or while watching a television program? If so, was it because you were touched emotionally by the situations of others? Did crying embarrass you, and did you attempt to conceal your feelings? Most importantly, did crying make you feel better?

Compassion is a balm for the soul. My sweet little wife Colene is easily touched by people who are hurting. She has an enormous compassion for children and elderly folks. Dozens of times while watching television, she has gotten choked up and laughed saying, "I think I'm going to cry." Her tenderness allows her to be touched when viewing starving children in Africa or someone else who is hurting. The response of empathetic tears is nothing less than genuine concern for others. This is beautiful in the eyes of a loving God who gave His only Son to touch mankind and bring eternal salvation.

I will never forget our first trip to the state mental hospital. Colene and I had gone there to give the patients of one ward a Christmas party, and the pitiful state of the patients broke Colene's heart and sent her out into the lobby sobbing.

When refreshments were served, one lady sat

crying because she couldn't get the lid off her ice cream cup. Another one was sobbing because Santa Claus hadn't arrived.

There was one patient I will never forget. She was a beautiful woman, no more than thirty years of age. Desperation and loneliness clouded her countenance, engulfing her soul in their paralyzing clutches. She sat rocking in her chair, all alone in her sorrow which was revealed by glistening tears streaming down her face like tiny rivers ebbing their way to the floor. It was apparent that this patron of misery felt unloved and untouched by the whole world.

Then something happened. Colene slipped quietly over to her and placed a small cheerfully-wrapped package into her pale quivering hands. For the first time in perhaps months, a smile etched its way across the unfamiliar territory of her face. She had been touched by someone who had reached out to her in love with a Christmas present. She had been tenderly touched by a total stranger who cared enough to give of herself to one of life's less fortunates.

The Bible declares that *love* is the most important attribute a person can possess (I Corinthians 13). Compassion is love in action. Jesus taught that the entire Ten Commandments given to Moses on Mt. Sinai could be expressed in two lone statements. "Thou shalt love the Lord thy God with all thy heart, and with all thy soul, and with all thy mind. This is the first and greatest commandment, And the second is like unto it, Thou shalt love thy neighbor as thyself" (Matthew 22:37-39).

Compassion is the tender feeling of brotherly love one has for his fellow man. This is the compelling force which directs the fortunate to help the less-fortunate. It directs the well-fed to feed the hungry — the clothed to dress the naked and the free to visit the imprisoned. It causes the well to care for the ill, the living to comfort the dying, the sheltered to provide for the homeless and the redeemed to reach the lost with the life-giving Gospel.

Nowhere in the scriptures are doctrines other than compassion and tenderness taught. Perhaps the best example is in Jesus' own Sermon on the Mount.

"Ye have heard that it hath been said, An eye for an eye, and a tooth for a tooth: But I say unto you, That ye resist not evil: but whosoever shall smite thee on thy right cheek, turn to him the other also. And if any man will sue thee at law, and take away thy coat, let him have thy cloak also. And whosoever shall compel thee to go a mile, go with him twain. Give to him that asketh thee, and from him that would borrow of thee turn not thou away.

"Ye have heard that it hath been said, Thou shalt love thy neighbor, and hate thine enemy. But I say unto you, Love your enemies, bless them that curse you, do good to them that hate you, and pray for them which despitefully use you, and persecute you; That ye may be the children of your Father which is in heaven: for he maketh his sun to rise on the evil and on the good, and sendeth rain on the just and on the unjust. For if ye love them which love you, what reward have ye? do not even the publicans the same? And if ye salute your brethren only, what do ye more than others? do not even the publicans so? Be ye therefore perfect, even as your Father which is in heaven is perfect" (Matthew 5:38-48).

I could share countless stories which demonstrate the effectiveness of the tender touch in regard to soul-winning. Many hardened and cynical spiritual desperados have had their stony hearts pulverized by the tears of someone who cared enough about their eternal destinies to weep with compassion or give them a gentle hug or handclasp. It is emotionally difficult to combat the tears and tenderness of someone who loves us enough to weep for us. Just as a soft answer turns away wrath, a tender touch can melt a hardened heart. Tenderness is the most important quality needed to effectively reach others with the Good News of redemption through the shed blood of Jesus.

Even on a strictly secular plane, tenderness is the greatest of attributes. Love is expressed in gentle, tender tones whether communicated by physical expressions or spoken words. An infant becomes calm as he responds to the gentle manner of his mother, and the tender touch of a husband serves to reassure his wife of his love for her. Wherever love can be found, the tender touch is also present. When we express our feelings, it is impossible to separate love and tenderness.

My father had a dry sense of humor and a gruff voice which could easily be misinterpreted if taken seriously. When teasing with Colene, who is a very gentle person, he would always crack a smile to assure her that he was kidding. Because of his love for her, he was tender even when appearing gruff. This is a fitting demonstration of the kind of tenderness needed as a companion to love.

A statement of love would seem futile without tenderness. How would a bride feel if, when the minister asked her groom, "Do you promise to love this woman until death do you part?" he grumbled, "Yeah, I guess so."

Saying "I love you" in a harsh tone will make an infant cry, because the gentleness has been removed from the voice inflections. The message we communicate through our gestures, body language and voice inflections is what others interpret us to say. Since our actions speak much louder than our words, without tenderness we will convey a wrong message.

It is most difficult to be kind to those who hurt us or to those we hold in contempt. Still, the

Christian virtue of gentleness is to be instituted at all times with all people regardless of past experience or personality clashes. We are not allowed the luxury of selecting the ones to whom our kindness will be shown.

God has never been selective in showing His love. He never once said, "I'll love *this* person because he has been reverent toward me; but I'll hate *that* person because he does not accept me." No! God gave his only begotten Son that whatsoever would believe upon Him should not perish but have everlasting life (John 3:16).

By so doing, God offered His ultimate expression of love and extended His personal tender touch to every person who has lived since Jesus died upon the Cross. The term *Christian* connotes being followers of Christ, striving to imitate Him in thoughts, actions and deeds. He expressed tenderness more fully than any person who ever lived and we who call ourselves Christians are mandated to follow His perfect example to the best of our abilities.

Jesus reached out to tenderly touch, console and bring comfort to the bereaved. When Mary and Martha were griefstricken over the death of their brother Lazarus, Jesus touched him with words of resurrection and brought comfort to the sorrowing sisters (John 11).

One day a lady suffering from a blood disease which had robbed her of her money and health, touched Jesus in a desperate attempt to be restored physically. In return for her faith, Jesus tenderly rendered healing for her body (Matthew 9:20).

Since Jesus extended His tender touch so often, I have listed other examples of His touching miracles in a following chapter titled "Tender Touching Miracles." You will note as you read that each miracle listed was performed because of Jesus' compassion for those who had need of His miraculous tender touch.

While Jesus' entire life was devoted to touching others, perhaps the most beautiful examples are found in the scriptural accounts of the crucifixion. Even amidst the agonizing tortures of death, His concern was not for Himself but for others. I would like to conclude this chapter with excerpts from The Living Bible accounts of John and Luke.

"When Jesus saw his mother standing there beside me, his close friend, he said to her, 'He is your son.' And to me he said, 'She is your mother!' And from then on I took her into my home.

"One of the criminals hanging beside him scoffed, 'So you're the Messiah, are you? Prove it by saving yourself — and us, too, while you're at it!' But the other criminal protested. 'Don't you even fear God when you are dying? We deserve to die for our evil deeds, but this man hasn't done one thing wrong.' Then he said, 'Jesus, remember me when you come into your Kingdom.' And Jesus replied, 'Today you will be with me in paradise. This is a solemn promise' " (Luke 23:39-43).

What a beautiful last scene! Jesus was still tenderly touching others, even during the perilous moments on the cross as He was drawing His last breath of life. What better example or model could we have for own lives?

If we refuse to extend a tender touch to those around us, we reject the major premise of Jesus' teaching and example. How can Christians be Christlike without a compassionate touch?

12

THE INTIMATE TOUCH

Men and women are created to respond instinctively to the physical touch which is a built-in factor for their natural expression of ultimate love. God designed this responsive touch to enhance the propagation of the world. In the Garden of Eden, He told Adam and Eve to multiply and replenish the earth. To accomplish this, He gave them each a natural sex drive which was a pure, sanctified instrument to be confined within the state of godly union known as marriage.

God told Adam, "Therefore shall a man leave his father, and his mother, and shall cleave unto his wife: and they shall be one flesh" (Genesis 2:24). Sex was ordained to be the most beautiful sacred relationship, joining man and woman in a physical, emotional and spiritual union.

The husband is only one-half of the conjugal union while the wife is the other half. In this sense they are not two, but one — and one-half plus one-half equals one, or my mathematics is incorrect. This mystical union takes two individuals, each with unique characteristics different from those of any other person who has ever lived, and joins them as one. Each maintains an individual identity while forming a cooperative personality in conjunction with their spouse. The "divine" matrimonial concept requires a blending of two people with mutual love into a singularity of purpose and cohabitation.

The act of marriage should never be taken lightly or entered into halfheartedly. Today's prevailing philosophy is, "Let's try it — and if things don't work out, we can always get a divorce." I ask, 'Does God recognize such an act as holy matrimony?' " My personal conviction is that any so-called marriage which exists for reasons other than love — such as convenience, money, sex, etc., is nothing less than legalized prostitution. Beloved, a marriage license does not make a marriage any more than owning a Bible makes one a believer!

My intentions are not to be harsh or critical, but to point out God's expectations for "holy matrimony" which is figurative of God's union with the church. Let's read what God says about matrimony: "Marriage is honourable in all, and the bed undefiled ..." (Hebrews 13:4). And, "Whoso findeth a wife findeth a good thing, and obtaineth favour of the Lord" (Proverbs 18:22). And again, "I will therefore that the younger women marry, bear children, guide the house, give none occasion to the adversary to speak reproachfully" (I Timothy 5:14), and "Husbands, love your wives, even as Christ also loved the church, and gave himself for it" (Ephesians 5:25).

A Godgiven mate will bind up your wounds with love's ointment, brushing away all hurts with a tender, loving touch. Your weakness will find supportive pillars of strength in your mate's understanding. Tears will be kissed away by the special one who stands by your side through riches and poverty, for better or worse — and most importantly, through sickness and health until death's separation.

Death is the final curtain for *holy* matrimony, while separation or divorce often writes the final chapter for secular matrimonial commitments. That which God constructs is designed to last forever, as well as to bring satisfaction, joy, bliss and contentment. The human body was built to never wear out — a fact which baffles the medical experts. They wonder how such a magnificent specimen of technology as the mortal body can develop malfunctions which result in death. The answer is simple. A body designed to live eternally, was cursed by God with a pronouncement of death as a result of man's sin against his Creator. Had Adam and Eve not succumbed to Satan's temptations, they and their descendants could have escaped death's angry clutches and lived forever.

Death came as a result of perverting God's perfect plan. Likewise, when people choose to do their own thing and go their own way void of God's direction, they should not be surprised at disastrous results. This is why so many people make improper choices in selecting their mates. Man, in his state of natural rebellion, often elects to go his own way, making choices apart from divine guidance. Next to a decision to accept Jesus Christ as one's personal Savior, the selection of a marriage partner is the most crucial choice any person will ever make. Why then, should such a decisive matter be void of God's supernatural influence?

While God designed man with the capacity to love and protect a woman, He also created him with the need for a helpmeet and companion to satisfy his emotional and sexual desires. Man's counterpart, woman, likewise was created with the need for a lover, protector and friend. God balanced the two sexes emotionally to complement each other's innate needs. Again, the two were designed to become one in body, mind, spirit and common purpose.

Man, by natural instinct, performs the role of the aggressor in lovemaking, while the wife's instincts prompt her to respond to his needs. Thus the needs of each are combined to bring mutual fulfillment as the two become one in the marital union. In the role of the aggressor, man has a natural desire to touch his wife, while her natural desire is to respond.

In sexual foreplay, women have a greater need to be touched, cuddled and caressed than do men. They are like fragile orchids which demand tender love and care in order to flourish. The husband is able to fulfill this need through the security of his strong loving arms as they yield compassion and tenderness.

It has been said there are fewer frigid wives than clumsy husbands. Perhaps this reflects the inability of many men to understand the principles of tender touching. I have counseled with scores of women who say they felt like inanimate sex objects to their impulsive husbands who failed to culminate loving relationships in lovemaking. These frustrated wives had two common complaints: the lack of touching prior to sexual encounter and the lack of cuddling and touching afterward.

A wife needs to be caressed after lovemaking to reinforce the fact that she is loved as a person and is not simply an object capable to supplying sexual gratification. Lovemaking followed by a quick goodnight kiss as the mate turns over to go to sleep conveys the message, "I got what I wanted, now I'm finished with you." This instead should be a tender, intimate reassuring moment of holding, touching and romantically caressing the one you love.

Men have a stronger sex drive than women which demands less stimulation by physical contact. Men are often sexually aroused by the sense of sight, while women require the sense of touch to reach the same erotic level. When men fail to realize this basic difference, sexual frustration often develops in even the best of marriages. Wives need to be needed as much as husbands need to be sexually satisfied. In God-centered marriages both can be accomplished, yielding multiplied blessings. Christian love dictates the necessity of putting the needs of others before one's own personal desires. In the marriage relationship, this means the husband tries harder to please his wife than himself, and the wife seeks to put her husband's pleasure before her own. This lifestyle should be carried to the bedroom where the most intimate expressions of love are shared.

Romantic feelings are not sinful, but are expected by God of those who have entered into the holy state of matrimony which God himself instituted. Christians have a responsibility of putting a little "spice" into their marriages instead of letting relationships die upon the vines of indifference. A man's handsome body was created for his wife's pleasure, as was a woman's beautiful body intended for her husband to enjoy. Being sensuous in one's own bedroom with a spouse is not unchristian, but emotionally wholesome. Moreover, temptations will be lessened for those whose sexual drives find gratification at home with the one they love.

When a mate does not want their body to be looked at, touched, kissed, or given to the other mate's enjoyment, that marriage is in trouble. Natural physical responses between husbands and wives are detailed in the Songs of Solomon.

This wise counselor describes the tender feelings of a man for his wife and her returned affections in his writings as a lover's discourse. The bride, bubbling with love's enchantments, happily croons to her husband, and he responds with words of romantic eloquence in the *Most Beautiful of Songs*, by Solomon (Good News Bible).

The Woman
2   Your lips cover me with kisses;
        your love is better than wine.
3   There is a fragrance about you;
        the sound of your name recalls it.
     No woman could keep from loving you.
4   Take me with you, and we'll run away;
        be my king and take me to your room.
     We will be happy together, drink deep,
        and lose ourselves in love.
     No wonder all women love you!
                        (Song of Solomon 1:2-4)

The Man
8   Don't you know the place,
          loveliest of women?
     Go and follow the flock; find pasture for
          your goats near the tents of the
          shepherds.
9   You, my love, excite men as a mare excites
     the stallions of Pharaoh's chariots.
10   Your hair is beautiful upon your cheeks
     and falls along your neck like jewels.
11   But we will make for you a chain of gold
     with ornaments of silver.

The Woman
12   My king was lying on his couch,
     and my perfume filled the air with
          fragrance.
13   My lover has the scent of myrrh
     as he lies upon my breasts.
14   My lover is like the wild flowers
     that bloom in the vineyards at Engedi.

The Man
15   How beautiful you are, my love;
     how your eyes shine with love!
                              (Song of Solomon 1:8-15)

The Woman
3   Like an apple tree among the trees of the
          forest,
     so is my dearest compared to other men.

I love to sit in its shadow, and its fruit is
    sweet to my taste.
4   He brought me to his banquet hall
    and raised the banner of love over me.
5   Restore my strength with raisins
    and refresh me with apples!
    I am weak from passion.
6   His left hand is under my head.
                  (Song of Solomon 2:3-6)

The Man
1   How beautiful you are, my love!
    How your eyes shine with love behind
       your veil.
    Your hair dances like a flock of goats
    bounding down the hills of Gilead.
2   Your teeth are as white as sheep
    that have just been shorn and washed.
    Not one of them is missing;
    they are all perfectly matched.
3   Your lips are like a scarlet ribbon;
    how lovely they are when you speak.
    Your cheeks glow behind your veil.
4   Your neck is like the tower of David,
    round and smooth,
    with a necklace like a thousand warrior
       shields hung around it.
5   Your breasts are like gazelles,
    twin deer feeding among lilies.
6   I will stay on the hill of myrrh, the hill of
       incense, until morning breezes blow.

and the darkness disappears.

7   How beautiful you are, my love;
       how perfect you are!

8   Come with me from the Lebanon
              Mountains; my bride;
       come with me from Lebanon.
       Come down from the top of Mount Amana,
       from the Mount Shenir and Mount Hermon,
       where the lions and leopard live.

9   The look in your eyes, my sweetheart
              and bride,
       and the necklace you are wearing have
              stolen my heart.

10   Your love delights me, my sweetheart and
              bride,
       Your love is better than wine;
       your perfume is more fragrant than any spice.

11   The taste of honey is on your lips,
              my darling;
       your tongue is milk and honey for me.
       Your clothing has all the fragrance
              of Lebanon.

12   My sweetheart, my bride, is a secret garden,
       a walled garden, a private spring;
       there the plants flourish.

13   They grow like an orchard of pomegranate
              trees and bear the finest fruits.
       There is no lack of henna and nard,
              of saffron, calamus, and cinnamon,
              or incense of every kind.

                          (Song of Solomon 4:1-13)

The Woman

10   My lover is handsome and strong;
        he is one in ten thousand.
11   His face is bronzed and smooth;
        his hair is wavy, black as a raven.
12   His eyes are as beautiful as doves by a
           flowing brook,
        doves washed in milk and standing by the
           stream.
13   His cheeks are as lovely as a garden that
        is full of herbs and spices.
      His lips are like lilies, wet with liquid myrhh.
14   His hands are well-formed, and he wears
        rings set with gems. His body is like
      smooth ivory, with sapphires set in it,
15   His thighs are columns of alabaster set
        in sockets of gold.
      He is majestic, like the Lebanon Mountains
        with their towering cedars.
16   His mouth is sweet to kiss;
        everything about him enchants me.
      This is what my lover is like,
        women of Jerusalem.
                        (Song of Solomon 5:10-16)

The Man

4   My love, you are as beautiful as Tirzah,
        as lovely as the city of Jerusalem,
        as breathtaking as these great cities.
5   Turn your eyes away from me;
        they are holding me captive.

Your hair dances like a flock of goats
   bounding down the hills of Gilead.
6   Your teeth are as white as a flock of sheep
   that have just been washed.
Not one of them is missing;
   they are all perfectly matched.
7   Your cheeks glow behind your veil.
8   Let the king have sixty queens,
   eighty concubines,
young women without number!
9   But I love only one,
   and she is as lovely as a dove.
She is her mother's only daughter,
   her mother's favorite child.
All women look at her and praise her;
   queens and concubines sing her praises.
10   Who is this whose glance is like the dawn?
She is beautiful and bright,
   as dazzling as the sun or the moon.
11   I have come down among the almond trees
   to see the young plants in the valley;
to see the new leaves on the vines
and the blossoms on the pomegranate
trees.
12   I am trembling; you have made me as eager
   for love as a chariot driver is for battle.

               (Song of Solomon 6:4-12)

The Woman
9   Then let the wine flow straight to my lover,
   flowing over his lips and teeth.

10   I belong of my lover, and he desires me.
11   Come, darling, let's go out to the
        countryside
      and spend the night in the villages.
12   We will get up early and look at the vines
        to see whether they've started to grow;
        whether the blossoms are opening
        and the pomegranate trees are in bloom.
      There I will give you my love.
13   You can smell the scent of mandrakes,
        and all the pleasant fruits are near our
        door.
      Darling, I have kept for you the old
        delights and the new.

                              (Song of Solomon 7: 9-13)

      I wish that you were my brother,
        that my mother had nursed you at her
        breast.
      Then, if I met you in the street,
        I could kiss you and no one would mind.
2   I would take you to my mother's house,
        where you could teach me love.
      I would give you spice wine,
        my pomegranate wine to drink.
3   Your left hand is under my head,
        and your right hand caresses me.
4   Promise me, women of Jerusalem,
        that you will not interrupt our love.

                              (Song of Solomon 8:1-4)

Beloved, God's Word plainly shows that romantic sensual feelings are wholesome when experienced within the confines of marriage. The preceding passages are the most beautiful descriptions of marital love in all of literature. This love discourse was inspired by the Holy Spirit who so aptly directed the recording of how a husband and wife should feel toward each other. I believe the Songs of Solomon were written as a blueprint for married couples. I believe it is to help them understand that the intimate touch is God's built-in mechanism whereby the sacred rite of conjugal love may find expression.

Isn't it wonderful to have a personal God who can meet all of our needs, whether physical, emotional or spiritual? We have a God who touches His servants on every level of life and death and allows us to personally touch Him through Jesus, our Mediator.

How sad to see individuals who are so bound by fear and inhibitions that they deny themselves much of the enjoyment of intimacy. As a counselor I have seen numerous cases — even among Christians — in which the wife timidly changes her clothes in the bathroom or closet, never allowing her husband to see her undressed. Sexual encounters are often mere duties and the playful touch is forbidden.

Beloved, this is not in line with God's plan for spouses! Paul teaches that when we enter into marriage, our bodies are no longer our own private property, but belong to our marriage partners.

This means that a husband has every right — legally and morally, to touch, to have and to behold the body of his wife. And she in return has the same rights to her husband's body. Paul tells us, ". . . to avoid fornication, let every man have his own wife, and let every woman have her own husband. Let the husband render unto the wife due benevolence: and likewise also the wife unto the husband. The wife hath not power of her own body, but the husband: and likewise also the husband hath not power of his own body, but the wife. Defraud ye not one the other, except it be with consent for a time, that ye may give yourselves to fasting and prayer; and come together again, that Satan tempt you not for your incontinency (I Corinthians 7:2-5 KJV).

The intimate touch is God's built-in mechanism designed to bring complete physical and emotional fulfillment between a man and woman whose lives have become one through the ordinance of holy matrimony. God intended that the conjugal touch should bring joy, as expressed in Proverbs 5:18. "Let thy fountain be blessed: and rejoice with the wife of thy youth." And again in Ecclesiastes 9:9, "Live joyfully with the wife whom thou lovest all the days of the life of thy vanity . . ."

I believe we could sum up what God is saying

in this way: "Husbands, reach out and touch your wives — and wives, reach out and touch your husbands!" By so doing, the chances for a lifetime of marital bliss will be greatly enhanced. God understood the importance of the intimate touch when he instructed Mark to tell all spouses, "What therefore God hath joined together, let not man put asunder" (Mark 10:9).

13

TEACHERS
CAN BE TOUCHERS!

Teachers have a unique opportunity for touching since they usually serve as educator, disciplinarian and role model to a child. Our cousin, Marge Bakker Klages, had worked for more than two decades with the Pasadena, California, school district as an elementary teacher. She has a remarkable understanding of human nature, as well as an extraordinary ability to communicate. When I told Marge that I had been asked by Dr. Hornbrook to co-author a book on the subject of touching, she responded by sharing the following three illustrations from her personal experience.

Tony was in my third-grade class. He was large, tough and cold. He was a handsome black boy. His strength used in hitting other children at times frightened me. Often I took him to the principal's office.

God gave me a special love for Tony. One day while sitting next to him, waiting for the principal, I put my cheek next to his and said, "Let me whisper something to you. You are such a beautiful boy inside. I like you very much, but I cannot let you hurt other people. Tony, I care," I assured him. "I am your friend. I know you feel angry. I feel angry too sometimes. That's when I want someone to care about me. I care about you," I continued, "and we're going to make it. I will never hurt you, I promise. You will have to be punished for fighting and hurting someone today, and be sent home. I'll see you tomorrow. Tony, I do care — and I love you."

We made it through that school year. I knew that Tony had come to regard me as not only his teacher, but his friend, as he handed me a picture taken in his Little League uniform. I look at it and recall his tears coming down on my cheek next to his and my arm on his shoulder. I remember his father having to go every day to be on a dialysis machine and regretting that he couldn't work and support his family.

I saw Tony recently. He is now in high school. He assured me he is doing fine. His smile and warm reception brought tears to my eyes again as I recalled the many, many times I had put my arm on his shoulder and whispered reassurance in his ear. Thank God for the privilege of being

His touch to a hurting child.

Joel is tall and handsome. His parents are professional musicians. Their divorce when Joel was in the third grade caused him extreme anger. He would break toys, his bicycle, and one time even a friend's leg. When Joel came to my sixth-grade class, he was somewhat "on trial" to see if he could cope with life in a public school classroom. His uncontrollable temper had, in the past, necessitated his being placed in a small private school.

At times I would see Joel's face redden, his fists clench and would watch his intense restraint. One day I quietly put my hand on his shoulder as I bent down to whisper in his ear, "Get out of this room. Get up quickly and quietly follow me."

On the school grounds are steps leading down a walkway and a lovely green slope nearby where we could sit. There was no one else around. With my hand on Joel's shoulder, I told him I admired his restraint to control his temper. And I told him it was all right now to cry, or to talk, or to walk around since we were alone in that campus area. He did cry. I assured him that each of us feels like crying at times, and that while he had exercised great strength in the classroom, he should now — in private — "cry it out."

Having a need to return to my classroom, I spoke to Joel. His shoulders were still heaving gently beneath the touch of my hand. "Don't ever be ashamed to cry or to share your feelings with a friend," I told him. "I want to be your friend — I care. Whenever you feel the need to leave the room in order to restrain your pattern of explosive temper, just quietly slip out and come here — I will know."

I gave him a Kleenex tissue and told him to return to the class when he felt he could. He did so in a few minutes. I knew he would. I trusted him. We were intimately honest. I was not judgmental, but affirmed his ability to control his temper. I affirmed his intensity of feeling and his supreme strength. Thank God for that priceless relationship, when again, His love and wisdom came through my touch.

There was a man retired from the telephone company who came to give his time when needed at our school. A boy in my class named Jonathan spent an hour by himself with this new "friend." Sometimes they talked — sometimes they played ball. But most of all they built a one-to-one relationship. I would see them sometimes — big man with small boy, with big hands on small shoulders, saying more than words could say, "You're all right. We are friends. I accept you."

I saw Jonathan's attitude mellow, his personality lose hostility, and his face begin to smile through that experience. That was third grade.

When I see Jonathan now, eight years later, he smiles and waves. Was a gift of time, friendship and touch worth one hour a week? I believe it changed a hostile boy!

My friend Shirley Sara operates a boys' home in Florida which includes a private Christian school. She shared with me how touching had become important in her relationship with the children in her charge. She had ordered some printing done, and when she picked it up, she noticed that the printer had included as a bonus, a stack of coupons reading, "Good For One Hug."

Shirley had won the confidence of the boys by introducing them to a stable, well-disciplined lifestyle which most of them lacked at home. But she had never made a practice of hugging the boys. For one thing, she felt some of the students who were nearing manhood would feel awkward or embarrassed by such closeness and she didn't want to make them uncomfortable.

But now that she had the coupons, she decided to go ahead and distribute them, then let each boy decide on his own when to "cash his coupon in."

The results were most surprising as during the next few days hands began springing up spontaneously throughout the small classroom. At each signal, Shirley would take the coupon and respond with the promised hug.

Finally, only one boy remained who had not cashed in his coupon. He was a sixteen-year-old youth named Don. He had had some tough things to deal with in life and seemed afraid to get close to anyone. But the right moment finally came, and Don cautiously slipped a hand up to signal Shirley's attention.

"I want a hug," he whispered, as she leaned close to speak to him. He stood up and they embraced. To Shirley's surprise, Don was just as reluctant to let go as he had been to be hugged in the first place.

The coupons had added a whole new dimension to Shirley's relationship with her students. Especially Don — the one who had seemed unreachable!

As proof of his trust in her, a short time later Don allowed Shirley to pray with him to accept Jesus Christ as his Savior.

14

MOMMY, PLEASE TOUCH ME

Johnny was a timid eighth-grader who could not get along with women teachers. After being suspended four times for causing classroom problems by arguing with female teachers, Johnny was expelled from school. But Johnny's problem was not that he harbored a vehement dislike for his teachers. It was instead that they represented an image of his mother whom he both hated and adored. It was these conflicting feelings which caused him to be expelled from school.

Whenever we would discuss his relationship with his mother, Johnny would talk about her like she was an angel seated upon a pinnacle of goodness. However, he would drop hints of his resentments toward his mother stemming from emotional traumas of the past. For a young man barely into puberty, Johnny had experienced more hurts than most of us undergo in a lifetime.

He had lived with a line of drunken stepfathers besides a host of live-in lovers with whom his mother had become involved during his grade school years. The times when he needed the loving touch of his mother, she was in the arms of someone she had picked up in one of the cheap bars where she sang or performed as an exotic dancer. She could touch strangers, but not the ones who so desperately needed her, the forgotten children left at home to fend for themselves.

The only memories of parental contact Johnny and his brothers and sisters had were of beatings by drunken stepfathers. Not only were they beaten and battered, but they had seen their mother assaulted many times. She was hospitalized on several occasions before the last stepfather was removed from the picture by another divorce. By this time, divorce had become a way of life offering temporary sanctuary until another drunken bum would marry their mother or until she would invite a live-in lover to share her home.

Amidst this humdrum of confusion, Johnny cried out for the love of a maternal touch which never came. His attitude toward life became cynical, evolving into deep-seated hostilities

which festered like a smoldering emotional vol-
cano, capable of erupting at any moment. He
became a walking human time bomb, looking
for a place to explode.

What would seem to be the answer to Johnny's
frustrations never materialized. When he was
eleven, his mother found Jesus and became a
born again believer. Instead of singing in bars
and nightclubs, she was singing praises to God in
churches. Her life of sin and blatant immorality
became past history, eradicated by the blood of
the Lamb. The old lifestyle had passed away,
being replaced by a moral code of high integrity.

This former good-time lady of the night became a dynamic witness for Jesus and became
active in a full gospel church. Her turn-around
was so dynamic, that people began coming to
her for spiritual advice and personal counsel.
She developed a reputation as a woman of un-
waivering faith and spiritual strength who was
able to touch strangers as well as friends. She
became a tremendous soul winner and witness
for Christ's transforming power.

Now comes the negative heartbreaking "but."
But she never touched Johnny. When it came to
touching her own child, she was still incapable
of displaying her emotions. Perhaps guilt from
past escapades restricted her from showing her
inner feelings because her children were aware
of her immoral past. Whatever limited her from
touching the ones she loved the most, they were

the ones who paid the price for the absence of her touch.

With a changed lifestyle, Johnny's mother was able to provide her children with food, clothing, spending money and a lovely home. But they never received what they needed most — the touch of a mother's love. Wealth and comfort can never suitably replace the touch of one who cares!

Johnny's mother, who I will call Sue, realized how much Johnny needed to be held and re-assured, but she never found the strength to overcome her hangups. She cried, prayed, pleaded, and sought God for an outpouring of supernatural love to help her show her inhibited feelings to her family.

Instead of walking in faith, expecting the love she prayed for to shine forth, Sue was guided by her feelings. The feelings of love never came, so she withdrew further, never touching Johnny or his brothers or sisters. All became losers. One child got involved with drugs, two children be-came incorrigible, and Johnny became a prob-lem child at school provoking his expulsion.

Because of her frustration over the problems of her children, Sue withdrew even further from her responsibility as a mother. She still could not bring herself to touch Johnny. In a desperate attempt to escape her misery, Sue tried to take her own life. When the suicide attempt failed, she became involved once again in an illicit affair which stripped her of her pride and Christian witness. She had yielded to her own desire to be physically touched while denying her children the right to a mother's touch. Unnecessary tragedy befell an entire family because of a mother's failure to respond to the plea, "Mommy, please touch me."

In case you're wondering how the story ends, it isn't completed as of this writing. Johnny has been assigned to an alternative school for problem youngsters where he is receiving extensive counseling. Sue has ended her immoral relationship with the man with whom she was involved. She has repented and is working through the difficult task of regaining her self-respect and her Christian testimony. She too is receiving counseling in an attempt to overcome her "hands off" policy with Johnny and the other children. Perhaps one day in the near future she will not turn deaf ears to the pleas of her children to give them a motherly hug.

Beloved, James tells us, "Therefore to him that knoweth to do good, and doeth it not, to him it is sin" (James 4:17). Please permit me the privilege of paraphrasing what the apostle James said, and apply it to touching. I believe God would tell us, "To him that knoweth the importance of touching others, and refuses to touch them, to him it is sin."

Please, please, please reach out and touch those around you and especially those in your family. Let it never be said that you failed to give to your loved ones the most important gift of all; your loving touch.

15

THE PARENTAL TOUCH

Physicians and psychiatrists have discovered that newborn infants learn the touch and smell of their parents even before visual recognition exists. A contented, happy baby is one who receives the reassuring touch of the loving mother which is vital to his security and development. A child needs the cradling, loving arms and tender kisses far more than the "oohs" and ahs" or baby jibberish so characteristically associated with newborn babes.

Being touched does more for the new arrival than all the teddy bears, rattles, and baby talk combined. Even infants seem to say, "Don't just talk to or view me like a canary in a gilded cage; but please, oh please, reach out and touch me.

I'm not breakable. Show me your love by your touch."

Before his shiny eyes are able to focus, an infant will reach out to touch his mother's breast or to eagerly clutch an extended finger. One of the first actions little ones learn is to smile in response to the familiar loving touch of a parent.

In our culture it is the mother's role to care for the physical needs of the infants. However, if fathers would realize how important their touch is to the tactile needs of their infants, the mother's role in touching would likely change.

Tradition asserts males to be "macho" individuals, and anything which seems closely related to women's work is considered unmanly. This pseudo-hypothesis robs many fathers of multiple blessings. Fathers would benefit greatly by bathing, kissing, rocking, holding, diapering, carrying, feeding, and playing with their children. And the child would experience great tactile satisfaction while the father strengthened emotional bonds with the flesh of his flesh which he had learned to affectionately touch.

For the most part, fathers who establish tactile bonds with their children in infancy do not find them out of reach during the teen years when most youngsters feel they are too big to be hugged, kissed, and touched by their parents. Children never outgrow the need to be touched by the ones they loved first, and this close relationship can continue throughout adult life. Most children who are embarrassed by their parent's presence in public or the parent's desire to give them a goodby hug before going to school, were not adequately touched during the developmental years.

Children who are sufficiently touched by their parents are more open in affection. They are less timid and fearful of showing how they feel because lines of emotional communication have been kept open through the touch which supersedes oral expression. Touched children are more secure because they receive physical reinforcement proving the parent's love.

Parents tend to ignore touching, hugging, and kissing their children as they grow old. What a tragedy! Children who have become accustomed to physical affection and no longer receive this tactile support can become insecure, confused, and emotionally distressed. In their minds, the question arises, "Why do my parents no longer hug me? Don't they still love me, or have I done something wrong?" Children may never ask these questions of their parents, but they wonder why they are no longer touched by the ones who used to show them so much affection and physical attention.

Bumper stickers across the nation ask an imperative question. *"Have you hugged your kid today?"* I believe parents should hug their children *every* day and tell their sons and daughters, "I love you." A goodnight kiss is the best gift any parent can give a child. Jesus said, "Or what man is there of you, whom if his son ask bread, will he give him a stone? or if he ask a fish, will he give him a serpent?" (Matthew 7:9-10). No parent would think of giving a hungry child a stone or snake. Yet many love-starved children go untouched and emotionally ignored by parents who unthinkingly give their children the brush-off when they need attention, and a "see you later" when they desperately need a hug. Parents forbid not your children to come unto you! Hug them! Touch them!

Dear friend, your touch is coveted by those who cherish and respect you. Having worked in the school behavioral field for nearly two decades, I could write a large book on the kids I have seen misbehave merely to attract attention. Society, for the most part, is a very unpersonal atmosphere, void of the loving touch. School consolidations for monetary and academic upgrading have created vast education mills. Timid students are often lost in the shuffle, trading personal identity for a number in the teacher's gradebook. An identity crisis can lead to a dehumanization of individual feelings of self-worth. Humans have feelings; numbers merely specify.

Most likely, you are either a parent or a potential parent, so please permit me to share with you the importance of touching your children. If you do not reach out to your own, they will seek out others who will pay special attention to them. These are the kids that homosexuals and sexual deviates prey on. Either will readily admit they don't stand a chance of making advances toward children who are shown the proper affection at home. It is the "untouched" who fall victim to their sin-warped webs of emotional entanglement. An innocent child can become marred for life because he or she yielded to a simple desire to be touched which was accommodated by the wrong person.

Take the time to take a long, good look at the real needs of your children. I am not talking about the physical provisions which can be stockpiled, but the emotional yearnings waiting to be fulfilled. Children need to be hugged, kissed, patted on the back, praised when deserving, and disciplined when they misbehave. Being a responsible, sensitive parent is the highest calling before heaven and probably the most difficult. It takes combining the patience of Job, the wisdom of Solomon, the strength of Samson, the leadership of Moses, the tenderness of Mary, the valor of David, and the guidance of God Almighty! Parenting goes far beyond bringing children into a troubled world. It includes guiding them through this life suitably making preparation for the life to come.

The next time you hear the "cloppity clop" of rapidly approaching footsteps, and the banging front door, before you scream, "Don't bang the door!" see what your child is so excited about. It might be a fine report card which motivated that dynamic entrance. The need to be praised might have impelled that youngster like a missile into your quiet domain, shattering the air with enthusiasim. Never forget, children .exert far more energy than adults. Also, children display emotions more openly since they haven't learned to suppress them like "refined" adults.

Last week my eight-year-old son left his bike parked behind my car and out of vision from the driver's seat. Fortunately, I was backing slowly when I heard the scratch of metal against metal and had sufficient time to hit the brakes, avoiding serious damage. Immediately, the bike was parked, being grounded for one week. When I told Colene about the incident, I said, "That boy has got to learn responsibility, because he sure is irresponsible!" "He sure is," she laughed, "he acts like an eight-year-old!" Even we so-called "experts" sometimes forget that children are only children. If they weren't, they would be adults! Children are not capable of responding to situations in an adult fashion. Their level of maturity cannot be expected to be that of a fully-developed human being.

Too often, unreasonable expectations are placed upon children by well-meaning parents, and the result is frustration for the youngsters. Each child is completely different emotionally, even in the cases of identical twins. Two siblings may look or act alike, but a look beneath the surface will reveal two totally different individuals. Younger children are often expected to be like an older brother or sister which serves as the "model." Millions of frustrated children attempt to fill the shoes of an older brother or sister, which is an unfair position in which to be placed.

Going back to the importance of the touch, firstborn children usually develop more rapidly than latter birthed. First children traditionally get more attention, holding, kissing, cuddling, and are worked with more by the parents. They are the "apple of the parents' eyes," basking in the center arena of attention. Their accelerated development makes them appear smarter than their younger counterparts which is in no way true. They were emotionally fertilized while the younger siblings, at the same stage of development, had to share parental time, attention, and touching.

Studies have proven that the only child is touched more and given more parental attention than children with siblings. This is not to point an accusing finger at anyone, but to admonish parents to realize the importance of their love, time, and touch as being essential to the wholesome development of each of their children.

Psychology talks a lot about sibling rivalry, which is nothing more than youngsters scrambling for their parents' attention and approval. Emotionally-neglected children will act up to get the attention they so badly need. They may abuse younger brothers or sisters as a way of getting the parents to notice them. To the emotionally hungry, negative attention seems more favorable than no attention at all. The innate need to be accepted, loved, touched, and noticed stimulates youngsters to achieve merely for rewards of parental praise.

It is extremely sad to see children "act up" because they are starved for touching. However illogical, emotionally-depraved children often become anti-social in an attempt to attract attention to themselves. They are willing to receive punishment in order to be noticed. One boy at the Alternative School where I work as a counselor has had three black eyes in the past six weeks. His need for attention is so great that he agitates others just to get noticed. Hostility does not motivate him. In fact, he will not fight; but chooses to become battered to receive contact from others. Children so desperately crave attention, they will get it through such negative contacts rather than go unnoticed.

A high percentage of today's children suffer from an "identity crisis." Kids with this syndrome feel like they are merely a number in life's vast computer. Devaluation of individual dignity is magnified by youngsters with low self-esteem. They feel inept and out of step with the rest of their generation. Generally, a feeling of *"no one understands me or wants to hold me and kiss away the hurts,"* is experienced by youngsters who are emotionally starved. What a difference one little hug could make! Those who are hurting are more than statistics, they are *real* people with *real* needs going unmet.

Children did not ask to be brought into this turbulent world. They arrive as a result of two people's actions, whether intentional or accidental. Kids deserve more than just being *born*. They have the right to be loved, cuddled, directed, and touched by their parents whether natural or adopted. In circumstances involving orphaning, this responsibility shifts to the legal guardians. Children should never be allowed to grow up like wildflowers or weeds. They should be gently cared for like rare precious orchids which demand a lot of attention.

Stable, structured homes are needed for the proper maturation of youngsters. With the divorce rate hitting five out of ten marriages in parts of the country, children are paying the price for the parent's failures. Hank Williams wrote of this plight when he penned the words relating a little girl's prayer. This heartbroken little darling prayed, "Mommy says Daddy has brought us to shame. No more am I to mention his name. Lord take me and lead me and hold to my hand. Heavenly Father, help me understand."

Children usually do not understand divorce.

They only know that one parent will be out of reach a majority of the time. Little children can't understand why mommy and daddy don't like each other anymore. They cannot comprehend why daddy or mommy isn't living with them. Little hearts are broken by the absence of good-night hugs and kisses. Divorce is cruel. It never considers the ones it hurts the most; children. Family circles are now being broken by divorce's cold wedge at an epidemic rate.

Single parents have an awesome responsibility which demands that they play the double role of both mother and father. This is certainly no easy task. The mother's tender touch should be received by children as well as the father's stern, corrective, loving touch. The traditional family unit, with God as the head, was the model designed by the Almighty. In His infinite wisdom, He established the guidelines which are essential to the proper raising of children. Pain results when His model is ignored. Substitutes for God's plans are always inferior to His perfect blueprint. Yet human nature, if not controlled by the Spirit, chooses second-best paths to trod when a better route was intended by the Master Builder.

Being a full-time parent is time-consuming as well as hard, and often frustrating, work. What I mean by a full-time parent is one who does not avoid parental responsibilities or difficult situations in parenting.

Today the parental dropout rate is swelling to epidemic proportions. Life's turmoil seems so great, parents are simply walking away from their families in order to avoid responsibilities associated with being mothers and fathers. When this occurs, everyone is hurt. The children feel dejected and the other parent is left alone to fulfill the double role of both parents.

God tells us that being a godly parent will be joyful and rewarding, but never once said it would always be an easy role to accept. His Word declares the exalted position of parenthood. Let's look at what His Word tells us in what I call "parenting principles."

Parenting Principles

Parental Joy
Psalm 127:4,5

As arrows are in the hand of mighty man; so are children of the youth. Happy is the man that hath his quiver full of them: they shall not be ashamed, but they shall speak with the enemies in the gate.

Parental Influence
Proverbs 22:6

Train up a child in the way he should go; and when he is old, he will not depart from it.

Parental Instruction
Isaiah 28:9

Whom shall he teach knowledge? and whom shall he make to understand doctrine? them that are weaned from the milk, and drawn from the breasts.

Parental Love
Titus 2:2

That they may teach the young women to be sober, to love their husbands, to love their children.

Parental Glory
Proverbs 17:6

Children's children are the crown of old men; and the glory of children are their fathers.

Parental Heritage
Psalms 127:3

Lo, children are a heritage of the Lord: and the fruit of the womb is his reward.

Parental Pride
Proverbs 15:20

A wise son maketh a glad father; but a foolish

man despiseth his mother.

Parental Promise
Acts 2:39

For the promise is unto you, and to your children, and to all that are afar off, even as many as the Lord our God shall call.

Parental Honor
Ephesians 6:2,3

Honor thy father and mother; which is the first commandment with promise; That it may be well with thee, and thou mayest live long on the earth.

Parental Obedience
Ephesians 6:1

Children, obey your parents in the Lord; for this is right.

Children need their parents to be available in times of need. When a youngster has a burning fever, skinned knees, or is being called foul names by the neighborhood bully, nothing is more reassuring than a parent's tender touch. Parents should serve as leaning posts or supportive pillars to their own flesh and blood. When a low grade is earned in school, an athletic contest lost, or any number of other failures have been experienced, children need to be supported by concerned parents who will put their arms around the discouraged child and tell him the war isn't over yet. One battle may have not been won, but victory can be achieved by persistence. Nothing can be sweeter to a defeated child's ears than hearing his parents say, "Together we will work this problem out, and together we will win."

The story of the Prodigal Son demonstrates that a parent is always a parent and a child is always the parent's child. Regardless of what a child does or how miserably he may fail, that child is still the flesh and blood of his or her parents. Even though acts committed by their children cannot be condoned, parents need to always stand by their children giving them moral support and spiritual guidance. A child who turns out to be a mammoth disappointment is no less a child than the one who is the epitome of success. Perhaps the failure was just one hug shy of success!

Drugs, pregnancy, crime, or social decadence should not separate a child from a parent's love. The God-kind of parents are those who will let nothing come between them and their love for their children. Just as God never gave up on His chosen people (children), parents are not to give up on their children. Instead, parents are to uphold their namesakes before the Lord in perpetual, fervent prayer.

Beloved, when you hold your child, you are shaping the future. How you handle your children will greatly determine what your child will be like in later life. The destiny of the universe could well be in your hands, as your child may be the one chosen for leadership positions decades from now.

Even Mary did not fully comprehend what role her son Jesus would play upon life's stage. Granted, only one child grew up to be the Messiah; but many have grown up to decide the fates of their generations. Libraries are filled with the exploits of great men who were once children abiding in their parent's care, being molded to take their place in history.

Before concluding this chapter, I feel compelled to share one story demonstrating the importance of a parent's loving touch, even at the newborn stage. One of the most moving displays of maternal love I have ever witnessed was that of a beautiful Christian lady in California. Her son was born with an open spine which demanded multiple corrective surgeries, traction, and months of hospitalization following birth. The doctor's prognosis was ultimate death as an almost absolute with a slight chance for survival. The specialist informed the mother that if the child survived, he would be reduced to little more than a vegetable. Because of the oxygen supply being cut off during surgery, brain damage was anticipated and the result was almost certain to be severe retardation.

This devout mother insisted on breast-feeding her impaired infant son against the advice of the doctors who thought she was being foolish. They told her that she was wasting her time and unnecessarily torturing herself. These medical experts felt the child would not benefit from her touch because the anticipated brain damage would leave him unable to understand or respond.

Driven by a supernatural love coupled with unrelenting faith, this patron of kindness insisted on talking to her little boy amidst his piles of bandages, life-sustaining tubes, and bodycast connected to traction pulleys. Unable to reach him over the imposing apparatus, she stood on a chair to reach him when it was feeding time. She didn't care that she had become the talk and spectacle of the hospital. Weeks turned into months, but she never missed the feeding, standing on the chair beside his bed. Throughout torturous months she could be found talking, touching, and breast-feeding this little broken bundle of medical hopelessness as she kept her post beside him.

Her calm loving voice and gentle touch must have given the little guy the will to fight and live. He underwent numerous surgeries to correct his physical deformity, always being reassured by his parents sitting by his side during convalescent recovery periods. They stayed with him, talking and touching him with love as they wiped his brow and cared for him. Thank God these parents understood the importance of the parental touch!

Today, although this boy wears braces, he attends a normal first-grade class. However, he *isn't* normal, but not as predicted by the doctors. He attends two *special* classes each school day. One class is *third-grade* math and the other *fifth-grade* reading! Love's touch can do what medical science can never attain. The former laughingstock of the hospital is now the proud mother of a gifted child because she touched what appeared to be a hopeless case, never allowing anyone to steal her dream or destroy her faith. She proclaims, "You will do for love what you would not do for anything else." Psychologically, emotionally, medically, and spiritually, no value can be attached to a loving touch. It is priceless!

16

TOUCHING
THE CONDEMNED

We live in a negative society which thrives on bad news. Headlines of crime, violence and tragedy seem to be the mainstay of today's news media.

It is impossible to read or hear such accounts without it evoking some kind of emotional response. Feelings of anger are thrust toward the guilty, while hearts go out in sympathy to the innocent victims.

Perhaps we need to stop and take a closer look at just who the victims are. Some are all too obvious — the innocent people who have fallen prey to extortionists, muggers, rapists and even murderers. There is no way our system of justice can make amends for what these have suffered.

But what about the families of the accused?
They are also innocent victims — victims of
circumstance who must both live with the knowledge and bear the stigma of what someone they
love has done.

Innocent children are often ostracized by their
peers because of a parent's incarceration. Or a
young mother may face constant discrimination
because she is the wife of a *convict*.

Or to go still a step further — what about the
party who confesses to his guilt? Does his guilt
deny him the right to love and forgiveness? If this
were true, we would all be in a hopeless situation because God's Word says "For all have
sinned and come short of the glory of God"
(Romans 3:23).

Jesus said when we visit the prisoner, we are in
fact visiting him! (Matthew 25:36-40) His concern was always for the less fortunate, and His
purpose was always to touch and forgive.

I am reminded of a television spot produced a
while back by the Assemblies of God. A young
driver has just struck a child in the street. His
anguish is apparent as he falls across the hood of
his car, angrily pounding it with his fists. Mur-
murs can be heard from the crowd which has
gathered, and one voice speaks with utter contempt, "I hope he *never* lives this down!" Then a
voice-over says, "God forgives . . . let Him." A
simple illustration, but what a vital message! No
matter what the offense, God is always waiting

to forgive.

If adverse situations are handled with love and prayful concern, it is possible to touch people who would close us out under ordinary circumstances.

I have a beautiful friend named Wanda who for years owned and operated a ladies' dress shop. One day when I stopped by the shop, Wanda remarked that she had just been through a very trying week. Her shop had been burglarized in the middle of the night on Monday, then robbed at midday the following Saturday. She went on to tell me about the latter incident.

She had reached to answer the telephone just as she saw two men entering her shop. The phone call distracted her just long enough for the men to gather up armloads of clothing and head for the door. Unable to stop the men, Wanda watched them speed off as she was telephoning the police.

Although she was terrified by what had happened, she was grateful for the kind Christian officer who responded and filed a report on the robbery. Because of prompt action on the part of the police, the two men were soon apprehended.

The officer's warning to the men was a witness to the power of God. "You can't do anything to hurt this lady," he told them, "because she's a Christian. If you try to get away, the Lord will show us where to find you."

The two men signed a confession, but the clothing had already been disposed of. Wanda's loss, uninsured because of a legality in her policy, was eighteen hundred dollars. She began asking friends to pray, not that her loss would be recovered, but that the men who robbed her would find Christ.

As she sat in the courtroom at the hearing some days later, she glanced up to see a black lady — a longtime Christian friend, taking the seat beside her. The meeting seemed to have taken them both by surprise. The realization of what had happened finally struck as the lady said, "It was my son who robbed your store."

"Do you think I could talk with him?" Wanda asked.

The woman left to get her son and moments later he appeared and slipped awkwardly into the seat next to Wanda.

"Jesus loves you so much," she told him, her eyes brimming with tears as she spoke. "He wants to forgive you. Won't you accept Him and let Him give you a new life?"

The man showed little emotion, but sat in silence as Wanda told him about Jesus' love.

After the hearing was over and the man was in police custody, it was learned that he had escaped from a Pennsylvania prison three years earlier and would have to be returned to finish serving time on a murder conviction.

A few weeks later when I stopped in at Wanda's to do some shopping, she told me she had some exciting news. The young man's mother had just dropped by to tell her that her son had accepted Jesus in prison.

"I'm just so happy!" Wanda beamed. "I had prayed so much for that man, and it's worth every cent I lost just to see him come to the Lord."

Who knows if the young man would have been open to accepting Jesus if he had not first seen Him in the courtroom that day in the life of the lady he had robbed?

In her book, *She Plays In God's Garden Now*, Dixie Artz shares how the Lord gave her a supernatural desire and ability to touch others during a time of deep personal bereavement.

Her daughter, Brenda, was brutally murdered just days before her seventh birthday. She had been missing for three days when her strangled body was found in a neighborhood house whose owners were away on vacation. Ironically, the person being charged with the murder was Bobby, a young boy Dixie had once prayed with to receive Christ.

Sleepless nights, plus the stark reality of what had taken place, left Dixie emotionally and physically drained. Yet she was able to find comfort in the assurance that little Brenda was now safe with Jesus where nothing could ever harm her again. With these things in her mind, the night before the funeral, her thoughts turned to Bobby and his family. She knew what the Lord wanted her to do.

Walking alone the short distance to Bobby's house, she rang the doorbell and soon stood face to face with his father. Her attitude showed that her mission was one of peace, and she was invited in to join the roomful of grieving relatives.

She asked if she might have prayer with the family and Bobby's father said he thought that would be good. Then she proceeded to pray, thanking God for His love and asking that His peace would comfort this family. She asked too that He would come into their lives and become a constant source of help and strength to them, just as He had been to her.

When she had finished praying, the boy's parents thanked her for coming. She was then able to return home, knowing she had been obedient in demonstrating God's forgiveness.

While many may face such situations of tragedy or danger, the opportunity for touching the condemned is still unlimited. Today there are many fine ministries devoted to reaching the men and women in our nation's prisons.

In order to get more of an inside look at what prison ministries are doing to reach men and women for Christ, I talked with Jeff Park,

Director of Prison Ministries for the PTL Television Network.

Jeff explained how the ministry recruits and trains volunteers not only for visitation work, but for correspondence in their "pen pal" program. The most exciting thing was seeing how the various components of the ministry interlink to accomplish life-changing miracles.

An excellent example of this was the case of a thirty-six-year-old man who I will call Bill. Bill had received a life sentence when he was only sixteen. Thanks to a satellite dish being placed in the prison by supporters of the PTL prison ministry, he was introduced to the gospel message.

Bill wrote and asked for a Bible and a study course, available free to all inmates through the prison outreach. Soon after, a pastor (one of 7,000 volunteers who work to assist Jeff) paid a personal visit to Bill in prison. As a result, he gave his life to Jesus.

The pastor was shocked to learn that since Bill had entered prison twenty years ago as a teenager, he was his *first* visitor! Bill will soon be eligible for parole. After he is released, there will be an "After Care" program available to provide him with a Christian support system as he faces the challenge of adjusting to the outside world — one very different from the one he left twenty years ago!

Bill is just one of more than one million men

and women who make up our nation's prison population. Roughly forty per cent of these are incarcerated in state and federal facilities, while the rest are housed in city and county jails.

Your commitment as a "pen pal" could provide the link between one of these prisoners and the outside world. Like the pastor who visited Bill — yours could be the "touch" that would open a waiting heart to accept Jesus Christ.

For years, the effectiveness of our penal system has met with much debate. A stint in prison is viewed by some authorities as a valuable "learning experience." Still others contend that many inmates — especially young offenders — will re-enter society worse off than ever, having learned all the savvy of the experts. What an opportunity Christians have to change the negative side of the picture!

Since Christian satellite programming played such a vital role in Bill's life, I asked Jeff how many prisons had this tool available. He explained that only sixteen dishes had been placed at the time, but their goal is to place dishes in two more prisons every month.

Do you want to visit for/with *Jesus* in prison? if you would like to accept this great challenge, check with your church to see how you can help in their prison ministry. If your church does not have a prison outreach, why not help Jeff meet the exciting challenge of bringing the gospel to prisoners every day, via satellite?*

Who knows how many more "Bills" there are — waiting for someone to care enough to bring them the Good News of the gospel!

*If you would like more information on the PTL prison ministry, write to: Jeff Park, PTL Prison Ministries, Charlotte, NC 28279.

17

NORMAN

While trying to recall some outstanding present-day illustrations of touching, this story which Mike Adkins tells came to my mind.

Ever since Mike's song "Adoration" was first introduced on radio, he has come more and more to the forefront of evangelical circles not only as a singer, but as a preacher. At first I credited this to his ability to balance profound spiritual truths with a quick wit and sense of humor. But I think Mike's widespread appeal stems even more from his accessibility and willingness to touch others.

Sometime back, I asked Mike to tell me a little about a man named Norman who is the subject of a song he has recorded by that title.

Having himself grown up and worked in the coal fields of Illinois, Mike well understood the dangers associated with coal mining. An explosion, a cave-in or some other mining disaster makes the headlines. Yet long after the newspaper has yellowed and crumbled, the survivors still live with the pain and loneliness that comes with the loss of a loved one. Such was the story of Norman.

He was considered the town "oddball." The younger generation knew little about him except that he was good for a laugh or perhaps a cruel practical joke. Shy and withdrawn, Norman seldom spoke. When he did, it was an instant replay of whatever had just been said to him.

The house Norman lived in hadn't seen a coat of paint in more than a decade. His filthy attire and grimy countenance attested to the fact that there was no one who mattered to him.

As a schoolboy, Mike, like the other youngsters in the small community, had regarded Norman as somewhat of an oddity. True, Mike had heard some talk about the man's strange behavior being linked to trauma he had experienced in early childhood. His father had lost his life in a mining accident and his death had so overwhelmed the six-year-old boy that he never recovered emotionally. But these facts about Norman's past had for the most part become obscured by time.

Years had passed when Mike one day returned to his old hometown with his wife and children. They bought a modest home and began settling into the community.

Mike had noticed with some curiosity an old abandoned-looking house across the street, and inquired as to who, if anyone, lived there. Imagine his surprise when he learned that Norman was not only still around, but was now his neighbor!

The first real contact came one day as Mike watched Norman, who made a meager living mowing lawns, struggling to get an old mower running. Casually crossing the street, Mike walked up to where Norman — now very frustrated — continued to yank at the rope attached to the old mower.

"Having trouble with your lawn mower?" Mike asked.

"Having trouble with your lawn mower?" Norman repeated. Mike worked to help him get the motor running, determined to somehow penetrate the wall which kept Norman a self-imposed prisoner.

One day when Mike spotted Norman sitting alone in a local coffee shop, he slipped into the seat across from him. The stench of his unbathed body and dirty clothing made the meeting unpleasant. But Mike braced himself to speak — knowing his own words would only be parroted back to him.

"Norman," he said with concern, "have you ever thought about your soul — and about eternity?"

The man's dark, hollow eyes shifted toward him momentarily. "You know," he said soberly, "I have been giving that some very serious consideration."

The communications barrier had been miraculously broken! Now Mike could proceed with the next step. The Lord began revealing to Mike that Norman's problem was a lack of love. He was to humble himself and began taking the man different places with him.

As the friendship developed, Norman grew more responsive, although measurable progress was sometimes slow. Mike continued to demonstrate God's love and Norman reciprocated by consenting to a shower and clean clothing.

Now, as Norman ventured out in public with Mike at his side, he felt protected from the cruel encounters of the past. And another thing was happening. His fresh-scrubbed image was giving him self-assurance — a trait he'd never had.

The time came when Mike knew something had to be done about the unsanitary conditions of the old house. And since the Lord had given him the concern, he knew he was elected!

The job, as a whole, seemed impossible. Yet by taking on one room at a time, Mike was able to clean and decorate, gradually transforming the depressing rooms into a warm, cheery atmosphere.

When the other rooms had been finished, Mike warily scanned the bathroom. It should have been declared a state of emergency! He wondered what color the old linoleum was beneath the buildup of sooty grime. The bathtub, unused for years, had become a storage place for useless junk until Mike had emptied it to allow Norman to shower. And the old porcelain lavatory was masked by layers of dark gray scum.

As he began cleaning, Mike surveyed the room again. It was unbelievable what a little elbow grease and a can of scouring powder had accomplished! But one job still remained. There, in stark contrast to the now-clean tub and lavatory, stood the bathroom stool. But Mike tackled the job — if somewhat reluctantly — with the same care he'd given to the rest of the room. "God," he muttered under he breath, as he lay on the floor wrestling to clean and remount the ugly fixture, "you sure know how to make a person humble!"

After some months, Mike began to see the fruit of his labor exhibited in Norman's behavior. He now assumed almost an air of pride as he strolled along at Mike's side, clean-scrubbed and dressed in a new suit and hat.

Mike finally gained Norman's confidence to where he felt free to bring his pastor along for a visit. As they shared together from the book of Romans, Norman simply, but sincerely, asked Jesus into his life.

One element of surprise was the discovery that behind Norman's expressionless face was a good mind. In fact, he soon became an avid reader!

He was perhaps more perceptive than people had given him credit for. He could only believe in a God whose love could first be demonstrated by those claiming to know Him. And God had brought Mike back to his hometown and placed him in the house across the street to see if he would be willing to touch an eccentric old man!

18

TOUCHING THE RELUCTANT

We can't always take things at face value, because people are not always saying what you think you hear. The person who seems to shrink from another's touch may be simply hiding his true feelings. Perhaps he has been betrayed in the past by those he trusted. If so, he has likely built a barrier about himself to keep people at a safe distance. I am reminded of an incident which took place some years ago.

I asked a co-worker how her neighbor's son was doing after I heard he'd been hospitalized in another city with a serious illness. She told me the boy's family had just been to see him and he was doing well — but they could not erase from their minds the plight of a little boy who shared the same room.

In contrast to the other young patients who excitedly awaited a visit from parents, wondering what new book or toy they would bring, one small boy seemed to be in a world all his own.

He lay there, oblivious to the other children, playing with a flexible drinking straw and talking to himself.

"I don't care if no one comes to see me," he said mechanically, "I don't care if no one ever comes."

Was he really telling those in the room that he preferred not to have his rest disturbed by his family and friends? Of course not! What he was actually saying was, *"It hurts to be alone. It's humiliating to have the kids here know that I'm so unimportant. I wonder if my family sometimes misses me?"*

This took place years ago, but I've often wondered what became of this child. Did someone ever come along and give him the touch he was crying out for? Did this lonely episode leave him with an abnormal fear of separation? Only God knows.

We need to listen with our hearts, as well as with our ears and ask for the Lord's discernment. Sometimes people resist touch because they are afraid of being misunderstood — or afraid that they will be embarrassed.

I am reminded of an incident which took place years ago while Norm and I were living in Chicago.

We had met and become close friends with a beautiful blind couple, Bill and Ada. We visited often and shared many meals together. More often than not, these were spur-of-the-moment get togethers.

Late one afternoon as I fried chicken for our evening meal, I realized I had more than twice what we would eat. I asked Norm if he would like to call Bill and Ada and offer to pick them up for dinner.

They hadn't fixed dinner yet and said they'd really love to come, but their friend Margaret was visiting them. Norm assured them that she was more than welcome to join us. But there was still an almost strange hesitancy. Then Bill explained that Margaret— who was also blind— was extremely timid about eating with sighted people. Furthermore, they were concerned that, because of her size, she would be unable to fit into the seat of our little Volkswagen.

But Norm was persistent and was able to persuade the three to come. About an hour later, he was ushering the three guests into our dining room.

Introductions were made, and Norm seated everyone as I brought the food from the kitchen and placed it on the table. As we sat down to eat, I learned why Margaret had been so reluctant to accept our invitation.

An early childhood accident had left her totally blind and partially deaf. And while she had been taught to do many things, no one had ever worked with her to help her master the use of eating utensils. This had become a great source of embarrassment to Margaret.

She confided that when she visited her married children, they always set a place for her in the kitchen, away from the rest of the family as they ate. During her adult years, she had become a severe diabetic and was grossly overweight. Eating had become even further complicated by the loss of most of her teeth.

A little friendly small talk put Margaret at ease with us. "Now Margaret, you just go ahead and eat however you want to," Bill quipped, "Ada and I promise not to even *look!*"

We all laughed at Bill's "blind humor" and Norm and I assured her we wouldn't look either. Once she realized that she need not feel intimidated, she was able to enjoy — for the first time in years — a meal with sighted friends!

After dinner, we retired to the living room where we listened with amazement as Margaret continued to share the tragedies and triumphs of her life. Years earlier she had been widowed with five small children. Still, she had been able to raise them all to adulthood doing all her own housework, cooking and laundry.

As I watched Margaret talk, I could not help being awed by the enormity of her body. She seemed to look the same, whether seated or standing. She reached down and brushed her hand across the front of her dress, as if to smooth out a wrinkle.

"You know," she said timidly, "because of my size, I'm not able to find things in the stores that will fit me . . . so I make all my own clothes."

I felt confused as I tried to imagine Margaret operating a sewing machine. Then she explained, "I sew everything by hand."

I looked again at the simple, but neat, cotton dress she was wearing and told her I was amazed at her ability to put together a garment without seeing. Noting my interest, she folded back the hem of her dress enough for me to see one of the seams — a succession of tiny and surprisingly-straight stitches.

Before Norm took our friends home, Margaret thanked us for including her in our invitation. Bill and Ada told us later what it had meant for her to be accepted enough to be asked to eat at the same table with friends — when her own family refused her such courtesy.

Our paths never again crossed, and as poor as Margaret's health was, it's doubtful she even lived much longer. Still, her courage and determination amid multiple handicaps, left a lasting impression on us. I was glad Norm had been discerning enough when he called, to see beyond her reluctance!

19

ONE WHO REACHED OUT TO ALL

Probably most people who have had the opportunity of hearing the wondrous story of Jesus, have visualized Him reaching out to the little children saying, "... Suffer the little children to come unto me, and forbid them not; for of such is the kingdom of God" (Mark 10:14). While this is a beautiful scene, Jesus spent His entire life reaching out. He touched others from before His birth until after His death (John 8:58). "Now wait a minute," some of you are saying. "How could anyone reach out to others before birth and after death?" Both are beyond the natural limitations of mortal man, but not of a supernatural God. A God who is omnipotent (all powerful) can do anything, with one exception.

The only thing He cannot do is remember forgiven sins which have been covered by the precious blood of Jesus. His word states, ". . . For I will forgive their iniquity, and I will remember their sin no more" (Jeremiah 31:34).

Please journey briefly with me through Jesus' earthly visitation. First of all, we must establish that God was Jesus' Father. Mary was a virgin until after the birth of Jesus, and Joseph, the man who married her, was merely a stepfather to the immaculately conceived Son of the Most High. The angel Gabriel announced to Joseph, Mary's fiance', ". . . Joseph, thou son of David, fear not to take unto thee Mary thy wife; for that which is conceived in her is of the Holy Ghost" (Matthew 1:20).

God's ultimate desire to touch man's eternal destiny was fulfilled through the birth of Jesus Christ, the Son of the true and living God. John expressed it well when he penned the most treasured verse of the entire Bible, "For God so loved the world (man), that he gave his only begotten Son, that whosoever believeth in him should not perish, but have everlasting life" (John 3:16). Through Jesus, man was provided a means by which he might approach God. Not only could man come unto God, but now he could actually touch God through the personage of Jesus. Up until this time, God had remained afar off, out of reach of man except through chosen mediators. Jesus became the

Eternal Mediator who would allow every person the privilege of direct access to God through His name.

The Bible is silent regarding Jesus' life from the age of twelve until he is about thirty which was the legal age for a man in that day and age to proclaim the Word of God. However, I am confident that even during those silent years, Jesus touched so many people by His lifestyle that volumes could be written. After all, John said, "And I suppose that if all the other events in Jesus' life were written, the whole world could hardly contain the books!" (John 21:25 TLB).

Jesus' earthly ministry began and ended with His arms and heart reaching out to touch all who would receive His tender loving embrace of majestic love. It is most difficult to condense the "touching" ministry of Jesus into a single chapter when a series of books could not do justice to the affects His touch had upon people with whom He came in contact. The lives of the great men of all ages combined could never equal the impact of one solitary man, Jesus Christ the Lord. He never sat upon an earthly throne or traveled more than a few miles from His hometown, but His touch has been felt around the world and His dominion will be above all power. He truly is the King of Kings and Lord of Lords. This position was not given, but earned because He loved others more than life. He gave His life as a ransom for all who would allow Him to touch their lives. He extended His touch to every man, woman, boy, and girl who would accept His life-changing, life-sustaining touch.

Rather than go into an extensive dialogue on Jesus' exemplary lifestyle which was personified by His reaching out to touch everyone regardless of sex, position, or background, I will instead outline some of His miracles. Each of these demonstrate His intense compassion for fallen man whom He desired to restore by a forgiving touch. His touch not only brought forgiveness, but brought eternal life to a suffering, dying, lonely, frustrated world. Jesus' purpose was not to receive glory from His miracles, but to touch people with needs who were hurting. His glorification came as a result of reaching out. These miracles could be labeled "tender touching miracles." To more fully understand and appreciate the significance of each miracle, please read each account in full. This will help you to comprehend how Jesus reached out to touch suffering humanity.

20

TENDER
TOUCHING MIRACLES

Nobleman's son (John 4:46)
Peter's mother-in-law (Matthew 8:14;
 Mark 1:31, Luke 4:38)
Leper cleansed (Matthew 8:3; Mark 1:41;
 Luke 5:18)
Paralytic (Matthew 9:2; Mark 2:3; Luke 5:18)
Impotent man (John 5:5)
Withered hand (Matthew 12:10; Mark 3:1;
 Luke 6:6)
Centurion's servant (Matthew 8:5; Luke 7:2)
Woman with issue of blood (Matthew 9:20;
 Mark 5:25; Luke 8:43)
Blind men (Matthew 9:27)
Deaf and dumb ((Mark 7:33)
Blind man (Mark 8:23)

Ten lepers (Luke 17:12)
Blind man (John 9:1)
Woman with spirit of infirmity (Luke 13:11)
Man with dropsy (Matthew 20:30)
Malchus (Luke 22:51)
Blind men (Matthew 20:20; Mark 10:46)
Demoniac in the synagogue (Mark 1:26;
    Luke 4:35)
Demoniac (Matthew 12:22; Luke 11:14)
Demoniacs of Gadara (Matthew 8:28; Mark 5:1;
    Luke 8:26)
Demoniac (Matthew 9:32)
Daughter of Syrophenician (Matthew 15:22;
    Mark 7:25)
Lunatic child (Matthew 7:14; Mark 9:26;
    Luke 9:37)
Widow's son (Luke 7:11)
Jairus' daughter (Matthew 9:18; Mark 5:47;
    Luke 8:41)
Lazarus (John 11)

Jesus set the perfect example for us to follow in our quest to reach others. We, as Christians must be like our Master, possessing a willingness to reach out to all regardless of their station in life. We must be willing to touch even "LIFE'S UNTOUCHABLES."

Beloved, someday we will each be required to give an account of our touching before the hosts of heaven. Those who have followed God's mandate to reach out to all will be rewarded for their faithfulness while those who failed to do so will be judged for their direct disobedience.

If you are ever in question whether you should touch a certain person, ask yourself one imperative question. Would *Jesus* bother to touch that person? If the answer is yes, as it always is, then reach out and touch that individual as an extension of Jesus' hands. Remember, yours and mine are the hands Jesus uses to touch humanity! Jesus' holy hands were never defiled by sin, yet they touched many sinners. He despised sin, but gave His life for the very ones who commit sin. He reached out to those who hated Him as well as to those who loved Him — all because of His supernatural love for humanity.

ABOUT THE AUTHOR

Dr. John R. Hornbrook writes from a background of varied experiences including criminal psychologist, police officer, teacher, minister, college professor, alternative education counselor and private practice as family-counseling specialist.

The ministry of Christian counseling and personal evangelism has burned in Dr. Hornbrook's soul. He has as his motto, "There is no saint without a past, and no sinner without a future, and there is hope for you because you are somebody special." Sharing this philosophy, his ministry has been entirely dedicated to reaching the lost, rehabilitating the wayward, and teaching submissiveness to the operations of the Holy Spirit.

Aside from his family counseling practice, Dr. Hornbrook serves as a rehabilitation counselor at an alternative school (for students due processed from the school corporation for infractions of various laws). He also directs a national book ministry for prisons which supplies free books to inmates.

Dr. Hornbrook has appeared on most of the major Christian TV talk shows, including Jim Bakker, 100 Huntley Street, Good Morning Chicago, Lester Sumrall and Praise The Lord (in Phoenix and Los Angeles). He teaches Family and Marriage Seminars and has been invited as a speaker at the White House Prayer Breakfast for senators.

He holds a B.A. degree with majors in Bible and Latin; a B.S. with majors in English and History; an M.A.T. with a major in Institutional Counseling; Ph.D. in Counseling Psychology; and a D.D. awarded for accomplishments in the extention of numerous youth programs and rehabilitation endeavors. He is presently a doctoral candidate for Doctor of Education Degree, specializing in Childhood Development, Nova University, Ft. Lauderdale, Florida.

Dorothy Fanberg Bakker is a freelance writer who lives in Charlotte, North Carolina.

MORE FAITH-BUILDING BOOKS
FROM HUNTINGTON HOUSE

America Betrayed, by Marlin Maddoux. This hard-hitting book exposes the forces in our country which seek to destroy the family, the schools and our values. This book details exactly how the news media manipulates your mind. Marlin Maddoux is the host of the popular, national radio talk show "Point of View."

A Reasonable Reason to Wait, by Jacob Aranza, is a frank definitive discussion on premarital sex — from the biblical viewpoint. God speaks specifically about premarital sex, according to the author. The Bible also provides a healing message for those who have already been sexually involved before marriage. This book is must reading for every young person — and also for parents — who really want to know the biblical truth on this important subject.

Backward Masking Unmasked, by Jacob Aranza. Rock 'n' roll music affects tens of millions of young people and adults in America and around the world. This music is laced with lyrics exalting drugs, the occult, immorality, homosexuality, violence and rebellion. But there is a more sinister danger in this music, according to the author. It's called "backward masking." Numerous rock groups employ this mind-influencing technique in their recordings. Teenagers by the millions—who spend hours each day listening to rock music—aren't even aware the messages are there. The author clearly exposes these dangers.

Backward Masking Unmasked, (cassette tape) by Jacob Aranza. Hear actual satanic messages and judge for yourself.

Beast, by Dan Betzer. This is the story of the rise to power of the future world dictator—the Antichrist. This novel plots a dark web of intrigue which begins with the suicide-death of Adolf Hitler who believed he had been chosen to be the world dictator. Yet, in his last days, he spoke of "the man who will come after me." Several decades later that man, Jacque Catroux, head of the European economic system, appears on the world scene. He had been born the day Hitler died, conceived by the seed of Lucifer himself. In articulate prose, the author describes the "disappearance" of the Christians from the earth; the horror and hopelessness which followed that event; and the bitter agony of life on earth after all moral and spiritual restraints are removed.

Devil Take The Youngest by Winkie Pratney. This book reveals the war on children that is being waged in America and the world today. Pratney, world-renowned author, teacher and conference speaker, says there is a spirit of Moloch loose in the land. The author relates distinct parallels between the ancient worship of Moloch— where little children were sacrified screaming into his burning fires — to the tragic killing and kidnapping of children today. This timely book says the war on children has its roots in the occult.

Globalism: America's Demise, by William Bowen, Jr. The Globalists—some of the most powerful people on earth—have plans to totally eliminate God, the family, and the United States as we know it today. Globalism is the vehicle the humanists are using to implement their secular humanistic philosophy to bring about their one-world government. The four goals of Globalism are *A ONE-WORLD GOVERNMENT *A NEW WORLD RELIGION *A NEW ECONOMIC SYSTEM *A NEW RACE OF PEOPLE FOR THE NEW WORLD ORDER. This book clearly alerts Christians to what the Globalists have planned for them.

God's Timetable for the 1980's, by David Webber. This book presents the end-time scenario as revealed in God's Word. It deals with a wide spectrum of subjects including the dangers of the New Age Movement, end-time weather changes, outer space, robots and biocomputers in prophecy. According to the author, the mysterious number 666 is occurring more and more frequently in world communications, banking and business. This number will one day polarize the computer code marks and identification numbering systems of the Antichrist, he says.

More Rock, Country & Backward Masking Unmasked by Jacob Aranza. Aranza's first book *Backward Masking Unmasked* was a national bestseller. It clearly exposed the backward satanic messages included in a lot of rock and roll music. Now, in the sequel, Aranza gives a great deal of new information on backward messages. Also, for the first time in Christian literature, he takes a hard look at the content, meaning and dangers of country music. "Rock, though filled with satanism, sex and drugs . . . has a hard time keeping up with the cheatin', drinkin' and one-night stands that continue to dominate country music," the author says.

Murdered Heiress . . . Living Witness, by Dr. Petti Wagner. The victim of a sinister kidnapping and murder plot, the Lord miraculously gave her life back to her. Dr. Wagner—heiress to a large fortune—was kidnapped, tortured, beaten, electrocuted and died. A doctor signed her death certificate, yet she lives today!

Natalie, The Miracle Child by Barry and Cathy Beaver. This is the heartwarming, inspirational story of little Natalie Beaver—God's miracle child —who was born with virtually no chance to live —until God intervened! When she was born her internal organs were outside her body. The doctors said she would never survive. Yet, God performed a miracle and Natalie is healed today. Now, as a pre-teen, she is a gifted singer and sings the praises of a miracle-working God.

Rest From the Quest, by Elissa Lindsey McClain. This is the candid account of a former New Ager who spent the first 29 years of her life in the New Age Movement, the occult and Eastern mysticism. This is an incredible inside look at what really goes on in the New Age Movement.

Take Him to the Streets, by Jonathan Gainsbrugh. Well-known author David Wilkerson says this book is "... immensely helpful..." and "... should be read..." by all Christians who yearn to win lost people to Christ, particulary through street ministry. Effective ministry techniques are detailed in this how-to book on street preaching. Carefully read and applied, this book will help you reach other people as you *Take Him to the Streets.*

The Agony of Deception, by Ron Rigsbee. This is the story of a young man who became a woman through surgery and now, through the grace of God, is a man again. Share this heartwarming story of a young man as he struggles through the deception of an altered lifestyle only to find hope and deliverance in the grace of God.

The Divine Connection, by Dr. Donald Whitaker. This is a Christian guide of life extension. It specifies biblical principles of how to feel better and live longer and shows you how to experience Divine health, a happier life, relief from stress, a better appearance, a healthier outlook on life, a zest for living and a sound emotional life.

The Hidden Dangers of the Rainbow, by Constance Cumbey. A national #1 bestseller, this is a vivid expose' of the New Age Movement which is dedicated to wiping out Christianity and establishing a one-world order. This movement—a vast network of tens of thousands of occultic and other organizations — meets the test of prophecy concerning the Antichrist.

The Hidden Dangers of the Rainbow Tape, by Constance Cumbey. Mrs. Cumbey, a trial lawyer from Detroit, Michigan, gives inside information on the New Age Movement in this teaching tape.

The Twisted Cross, by Joseph Carr. One of the most important works of our decade, *The Twisted Cross* clearly documents the occult and demonic influence on Adolf Hitler and the Third Reich which led to the Holocaust killing of more than six million Jews. The author even gives the specifics of the bizarre way in which Hitler actually became demon-possessed.

Who Will Rise Up? by Jed Smock. This is the incredible— and sometimes hilarious— story of Jed Smock, who with his wife Cindy, has preached the uncompromising gospel on the malls and lawns of hundreds of university campuses throughout this land. They have been mocked, rocked, stoned, mobbed, beaten, jailed, cursed and ridiculed by the students. Yet this former university professor and his wife have seen the miracle-working power of God transform thousands of lives on university campuses.

Yes, send me the following books:

_____ copy (copies) of **America Betrayed!** @ $5.95
_____ copy (copies) of **A Reasonable Reason To Wait** @ $4.95
_____ copy (copies) of **Backward Masking Unmasked** @ $4.95
_____ copy (copies) of **Backward Masking Unmasked Cassette Tape** @ $5.95
_____ copy (copies) of **Beast** @ $5.95
_____ copy (copies) of **Devil Take The Youngest** @ $6.95
_____ copy (copies) of **Globalism: America's Demise** @ $6.95
_____ copy (copies) of **God's Timetable For The 1980's** @ $5.95
_____ copy (copies) of **More Rock, Country & Backward Masking Unmasked** @ $5.95
_____ copy (copies) of **Murdered Heiress . . . Living Witness** @ $5.95
_____ copy (copies) of **Natalie** @ $4.95
_____ copy (copies) of **Rest From The Quest** @ $5.95
_____ copy (copies) of **Take Him to the Streets** @ $6.95
_____ copy (copies) of **The Agony Of Deception** @ $6.95
_____ copy (copies) of **The Divine Connection** @ $4.95
_____ copy (copies) of **The Hidden Dangers Of The Rainbow** @ $5.95
_____ copy (copies) of **The Hidden Dangers Of The Rainbow Seminar Tapes** @ $13.50
_____ copy (copies) of **The Miracle of Touching** @ $5.95
_____ copy (copies) of **The Twisted Cross** @ $7.95
_____ copy (copies) of **Who Will Rise Up?** @ $5.95

AT BOOKSTORES EVERYWHERE or order direct from: Huntington House, Inc., P.O. Box 53788, Lafayette, LA 70505.

Send check/money order or for faster service VISA/Mastercard orders call toll-free 1-800-572-8213. Add: Freight and handling $1.00 for the first book ordered, 50¢ for each additional book.

Enclosed is $ _____ including Postage.

Name _____

Address _____

City _____ State and Zip _____